About Island Press

Island Press is the only nonprofit organization in the United States whose principal purpose is the publication of books on environmental issues and natural resource management. We provide solutions-oriented information to professionals, public officials, business and community leaders, and concerned citizens who are shaping responses to environmental problems.

In 2006, Island Press celebrates its twenty-first anniversary as the leading provider of timely and practical books that take a multidisciplinary approach to critical environmental concerns. Our growing list of titles reflects our commitment to bringing the best of an expanding body of literature to the environmental community throughout North America and the world.

Support for Island Press is provided by the Agua Fund, The Geraldine R. Dodge Foundation, Doris Duke Charitable Foundation, The William and Flora Hewlett Foundation, Kendeda Sustainability Fund of the Tides Foundation, Forrest C. Lattner Foundation, The Henry Luce Foundation, The John D. and Catherine T. MacArthur Foundation, The Marisla Foundation, The Andrew W. Mellon Foundation, Gordon and Betty Moore Foundation, The Curtis and Edith Munson Foundation, Oak Foundation, The Overbrook Foundation, The David and Lucile Packard Foundation, The Winslow Foundation, and other generous donors.

The opinions expressed in this book are those of the author(s) and do not necessarily reflect the views of these foundations.

Science
MAGAZINE'S
State of the Planet

Science

MAGAZINE'S

State

of the

Planet

2006–2007

Edited by

Donald Kennedy

and the editors of *Science* magazine

◤ AMERICAN ASSOCIATION FOR
THE ADVANCEMENT OF SCIENCE

⬤ **ISLAND**PRESS

Washington · Covelo · London

ISSN 1559-1158

British Cataloguing-in-Publication data available.

Design by BookMatters

Manufactured in the United States of America

10   9   8   7   6   5   4   3   2   1

Contents

Acknowledgments   *xi*

Introduction   *1*
DONALD KENNEDY

LIVING RESOURCES

Life on a Human-Dominated Planet   *5*
DONALD KENNEDY

Human Population:
The Next Half Century
13
JOEL E. COHEN

Prospects for Biodiversity   22
MARTIN JENKINS

World Fisheries: The Next 50 Years   29
DANIEL PAULY, JACKIE ALDER,
ELENA BENNETT, VILLY CHRISTENSEN,
PETER TYEDMERS, AND REG WATSON

SCIENCE IN THE NEWS

Taking the Pulse
of Earth's Life-Support
Systems   9
ERIK STOKSTAD

Will Malthus Continue
to Be Wrong?   18
ERIK STOKSTAD

Global Analyses Reveal
Mammals Facing
Risk of Extinction   24
ERIK STOKSTAD

The Sturgeon's Last
Stand   30
RICHARD STONE

PHYSICAL RESOURCES

Preserving the Conditions of Life   39
DONALD KENNEDY

Tropical Soils and Food Security:
The Next 50 Years   49
MICHAEL A. STOCKING

Global Freshwater Resources:
Soft-Path Solutions
for the 21st Century   59
PETER H. GLEICK

Energy Resources and
Global Development   69
JEFFREY CHOW, RAYMOND J. KOPP,
AND PAUL R. PORTNEY

Global Air Quality and Pollution   79
HAJIME AKIMOTO

Modern Global Climate Change   88
THOMAS R. KARL AND KEVIN E. TRENBERTH

THE COMMONS

Managing Our Common Inheritance   101
DONALD KENNEDY

The Tragedy of the Commons   115
GARRETT HARDIN

The Struggle to Govern
the Commons   126
THOMAS DIETZ, ELINOR OSTROM,
AND PAUL C. STERN

Social Capital and the Collective
Management of Resources   142
JULES PRETTY

Wounding Earth's
Fragile Skin   53
JOCELYN KAISER

As the West Goes Dry   65
ROBERT F. SERVICE

Can the Developing World Skip
Petroleum?   76
GRETCHEN VOGEL

Counting the Cost of
London's Killer
Smog   81
RICHARD STONE

The Melting Snows of
Kilimanjaro   91
ROBERT IRION

Fish Moved by Warming
Waters   129
MASON INMAN

Hopes Grow for Hybrid Rice to
Feed Developing World   158
DENNIS NORMILE

A New Breed of Scientist-
Advocate Emerges   162
KATHRYN S. BROWN

How Hot Will the Greenhouse
World Be?   177
RICHARD A. KERR

The Carbon Conundrum   180
ROBERT F. SERVICE

Uncertain Science Underlies
New Mercury Standards   186
ERIK STOKSTAD

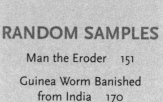

RANDOM SAMPLES

Man the Eroder   151

Guinea Worm Banished
from India   170

Managing Tragedies:
Understanding Conflict over
Common Pool Resources   149
WILLIAM M. ADAMS, DAN BROCKINGTON,
JANE DYSON, AND BHASKAR VIRA

Global Food Security:
Challenges and Policies   154
MARK W. ROSEGRANT AND SARAH A. CLINE

New Visions for Addressing Sustainability   161
ANTHONY J. MCMICHAEL, COLIN D. BUTLER, AND CARL FOLKE

The Burden of Chronic Disease   167
C. G. NICHOLAS MASCIE-TAYLOR AND ENAMUL KARIM

The Challenge of Long-Term Climate Change   172
K. HASSELMANN, M. LATIF, G. HOOSS, C. AZAR,
O. EDENHOFER, C. C. JAEGER, O. M. JOHANNESSEN,
C. KEMFERT, M. WELP, AND A. WOKAUN

Climate Change: The Political Situation   179
ROBERT T. WATSON

Tales from a Troubled Marriage:
Science and Law in Environmental Policy   183
OLIVER HOUCK

Index   195

Acknowledgments

Four editors of *Science*—Brooks Hanson, Andrew Sugden, Caroline Ash, and Jesse Smith—were centrally engaged with planning and executing the Special Issues that formed the basis for this book. The *Science* copyediting department handled the additional revisions for this book, and Pat Fisher helped in the preparation throughout. Our colleagues in News, especially Colin Norman, helped to identify News stories to lend illumination to each chapter and add current policy content. Alan Leshner, the Chief Executive Officer of the American Association for the Advancement of Science, along with its Board of Directors, lent enthusiastic support to this project, fulfilling the AAAS mission to support science *and* serve society. Todd Baldwin and his colleagues at Island Press provided splendid editorial help. Finally, the authors of all these chapters are members of a scientific community in which volunteer efforts to explain research and its relationship to human welfare are among its most admirable characteristics. We thank them for their willingness to contribute in this way.

Introduction

DONALD KENNEDY

The thoughtful reader will wonder, on opening this book, how this collection of explanatory essays got put together in the first place. Well, it was produced by a group of us who are the editors of *Science*, the weekly scientific journal of the American Association for the Advancement of Science—the world's largest organization of professional scientists. There are two dozen of us, all experienced experts in one or more particular disciplines of science. Most of the time, our task is to evaluate "papers"—manuscripts sent to us by scientists that report experiments and interpret the results of those experiments, or analyze observations of new phenomena, or formulate theories to explain conditions or events that are mysteries crying out for explanation. Because we get 12,000 of these manuscripts each year, that keeps us pretty busy.

Fortunately, we also get to spend time going to meetings and conferences where the best scientists in our individual specialties gather. It helps keep us on the cutting edge of where science is going. We also get to interact with a group of superb science journalists, who track what's new and exciting and explain it in the News and News Focus sections, which each week's issue of *Science* contains near the front of the magazine.

Fortunately too, our work yields us one other kind of opportunity, which has a lot to do with how this book came into being. A dozen or more of the 51 issues of *Science* each year are what we call "special issues"; several articles, reviews, and News stories are grouped into a single issue in order to give special attention to a particularly important scientific problem. Sometimes the problem is one that has a significance that reaches well beyond the science itself and contains broader implications for human societies and how they develop wise policies.

Two years ago we had become convinced as a group that the status of environmental resources—physical resources as well as living ones—was a problem to which science ought to pay increasing attention. Coming up, we realized, was an interesting anniversary. Thirty-five years before that time, Professor Garret Hardin of the University of California, Santa Barbara, had published an essay in *Science* called "The Tragedy of the Commons." More will be said about the Hardin essay later, but the essence of its message was this: The rate of human population increase was leading to an overuse of various resources that could not be sustained. Hardin argued that depletion would encourage each user to intensify the effort to exploit it, since the benefits would fall to the user whereas the costs of the action would be shared with all the others. That simple posing of the "commons problem" produced more interest and more requests for reprints than any other article the magazine had ever published.

What we know was that a more recent literature had developed in which new and different approaches to solving the commons problem were being tried by various societies—and being studied by social scientists. At the same time, we realized that the conditions under which commons resources were being harvested were changing, as human activities were altering everything from the landscape and the oceans to the climate itself. So we decided to do a special issue on the commons problem, reprinting the Hardin essay and following it up with more recent material.

Before we could get to that, however, our readers needed to have a picture of the status of global resources—beginning with an analysis of population growth and continuing through various pools of common-property resources—such as fisheries, terrestrial biodiversity, tropical soils, and so on. So before the scheduled appearance of the Hardin anniversary volume, we asked the most expert author we could find for each of the resources, and asked them to summarize the state of that resource for our readers. These were published, two papers in each issue, in a series we called "The State of the Planet." That title, of course, has found its way onto the cover of this book.

Finally, we have an obligation as your editors to explain why it's important for us to take your time with this material. You will find some of the reasons in the rest of this text. For example, every resource on this planet is subject in some way to the condition of its environment. The quality of fresh waters depends upon what happens to the forests in their watersheds. The success of agriculture depends upon the success of the surrounding native ecosystems that supply a variety of services: birds that are responsible for pest control and pollinators that enable crops to reproduce. The distribution of plants and animals in the wild is changing as the climate changes, and as humans introduce new species into regions where they are not native. You can name other dependencies of the same kind. To the editors of *Science*, these relationships—and the changes in them as humans continue to alter the world—compose the most important and challenging issues societies face. Without scientific understanding, those who will make policies in the future will be forced to do so without the most essential tool they could have.

Living Resources

Life on a Human-Dominated Planet

DONALD KENNEDY

Any account of the living resources on our planet must begin with an account of our own species. Those of us who lecture or teach about the status of natural resources have become accustomed to the first question that often follows such a talk: "Well, aren't you missing the point? Isn't the problem really the growth of the human population and its increasing call on resources of all kinds?"

The Influence of Humans

No species has ever achieved such total ecological dominance over so wide a range; indeed, every square mile of the planet is now distinctly influenced by humans. In treating the status of resources, especially slowly renewable ones such as biological resources, it's wise to begin with demography—the present and future of the human population. So we asked Joel Cohen, one of the world's most distinguished and thoughtful demographers, to set out the likely future in "Human Population: The Next Half Century." He tells us we will go from about 6.4 billion people to 9 billion before the end of the century we're in; that's the "median" projection from the United Nations and the World Bank. The new people will overwhelmingly be in the developing world, and increasingly they will live in cities. The numerical estimate could be lowered if AIDS and other emerging infectious diseases, such as new influenza strains, or expanding conflicts cause death rates to rise more rapidly then we expect. On the other hand, some developing nations are lowering their total fertility rates even faster than Cohen knew when his chapter was written. Thus there is still some guesswork in global population estimates.

But there is no guesswork about this: In many developing countries, history has bequeathed us a powerful legacy in what has been called "population momentum"— the growth force exerted because large percentages of their citizens are young. In some of the poorest nations, half the population may consist of individuals 14 and under. Most of these children are poor, and they are still rural. Their families need them for work: gathering wood and water, or cultivating. Others need them for other purposes, so some will become the child-soldiers all too familiar in African conflicts. If they mature, and many will, they continue to boost population numbers as they reach reproductive age. It is in such nations that total fertility rates are high—often far above replacement level.

In some of these nations, however, fertility rates are dropping faster than the predictions made in the late 1990s. What are the causes? Population programs developed and supported by international organizations and by national governments have been partly responsible, more in some cases than in others. The merging of AIDS control programs with efforts supporting reproductive health and population control has been of some help. But comparative data, especially from different provinces in India, suggest that the education of women and the correlated increase in the numbers of women who are employed outside the home are important factors in population limitation.

Much of what needs to be said on a cross-national basis about our population future depends on history. Look carefully at Cohen's graphs showing population size and age distributions for Europe and for Africa and West Asia. In Europe, total fertility rates are not only below replacement level now; they have been for some time. (It will surprise some that among the lowest rates are to those be found in countries with large Catholic majorities, like Italy and Spain.) The European nations display a bulge in their population profiles now centered in the late 30s; it is destined soon to become what could be called a "middle-age spread," and by 2050 the largest share of the population will be in their elder years. A profile for the United States would look somewhat similar.

That is why makers of social policy in the United States and Europe are seriously worried about how to handle economic planning about retirement. The Social Security system in the United States has depended on having more workers contributing to the system than there are retirees receiving payment. The economic systems supporting social welfare in Europe are even more stressed. Thus social planning for the next several decades is likely to produce significant social and political tension, as has been seen in the proposals for Social Security reform in the United States during 2005–6.

The problem is quite different in the developing nations. Even those that have achieved significant improvements in per capita income and have stable governments face a future that must accommodate an excess of young people. They will have labor-excess economies resulting in high unemployment rates, and they will have future population growth commitments that will demand careful resource planning.

Biological Diversity

That, of course, takes us to the next chapter. It is not an accident that the most rapid growth in the human population, and the greatest consequent pressure on natural resources, is occurring in the parts of the world in which biological diversity is highest. Martin Jenkins makes that point near the beginning of his account of the circumstances that conserving biodiversity will confront, in "Prospects for Biodiversity." There is little question that the addition of three and a half billion people will require an increase in agricultural productivity. Jenkins points out that using fairly optimistic productivity estimates, the addition of 20 million hectares of new agricultural land will be needed.

From the viewpoint of biodiversity, it matters a lot where the new croplands are located. Currently, for example, some of the largest gains in cropland are taking place in various regions of Brazil, and entail the loss of some of the most biodiverse forests in the world. Some increases in agricultural lands are occurring in places where plots of agricultural land are interspersed with areas of preserved forest, and there is much to be learned about the degree to which biodiversity is lost through the study of what Daily calls "countryside biogeography" (1). As agricultural productivity is increased, much of the effort will have to concentrate on improving the productivity of those lands that are already under cultivation—because too often, the alternative to increased yields on valley or lowland plots is that agriculture moves upland, cutting down forests in order to grow crops on more marginal soils. The result is a "double whammy"—lower productivity as well as loss of biodiversity.

The loss of biodiversity through species extinction is largely based on what is known about the relationship between species richness and the size of habitats: islands in some early studies, and isolated patches of forest in some newer ones. Scientists argue about rates of extinction, but there is little doubt that current rates are far higher than has been experienced in past "average" geological times. The recent extinction record is particularly clear for Pacific islands, where isolation has led to unusually large numbers of endemic species that are found nowhere else. The data are especially clear for birds, where lava tubes and other sites permit the identification of lost species through recent "fossils." It is estimated by Steadman that over 2000 species of birds, perhaps 20 percent of the world's identified species, have become extinct as a result of human occupation of these islands (2).

> The most rapid growth in the human population, and the greatest consequent pressure on natural resources, is occurring in the parts of the world in which biological diversity is highest.

Why should we care? A variety of arguments have been made, and they appeal to different groups. Many people have an affinity for the outdoors and wilderness, and it is important to them to feel that their explorations are taking them to habitats that are close to their natural state. Many of us underestimate the degree to which environments, even pristine-looking ones, have been changed by past human influence. Nevertheless, if environments look relatively undisturbed it makes a difference to those who are seeking a vision of nature. That is an important attribute of natural environments that are sought for particular purposes. Some of these are indirect—for example, many people want to contemplate the existence of a particular piece of conserved landscape even if they do not plan to visit it. Others are direct: people visit national parks and will pay to do so simply for the enjoyment they gain, but do not take anything away. That kind of direct use is called *non-consumptive* by economists, to contrast it with, say, trout fishing or duck hunting, in which the visitors are motivated by a harvesting purpose.

In each of these instances, biodiversity is important because it provides attributes that give an ecosystem its particular character and esthetic. They belong to a category of nature's values that have been called *ecosystem services*, a term used to cover a broad range of ways in which relatively undisturbed natural areas benefit humans. An important, relatively new movement seeks to attach economic value to some of these values, so that more convincing arguments can be made in the interest of conservation. Of course it's difficult to give a dollar value to the esthetic features of mature forests; but people do pay to visit them, and their travel costs and recreational purchases can be measured. Other examples: Farmers benefit from the pollination services provided by native bees and other insects. Crop losses occur if these are unavailable, and the services then have to be provided—at a cost—by supplying introduced bees and paying the beekeepers who tend them. Distant forests around reservoirs provide purification services for municipal water supplies, and if these are lost through development the city will have to install expensive treatment technologies to achieve the same result. It may cost the city less to buy out the developers than to install the new equipment, as was the case with New York City's reservoirs in the Catskill Mountains.[3] It is difficult to estimate the value of these ecosystem services on a worldwide basis, but guesses reach into the trillions of dollars. Not all environmentalists favor putting their feelings about nature into dollar terms, but the concept has made planners and policymakers think in a different way about the usefulness of conservation.

Marine Resources

Marine resources, especially fisheries, are under at least as much pressure as terrestrial ecosystems. In scoping out the "World Fisheries: The Next 50 Years," Daniel Pauly and a team of colleagues have identified the rates and causes of overharvesting. When open-ocean fisheries were developing in the 19th and early 20th centuries, the dominant view was that these fisheries were essentially inexhaustible. There is something about the opacity of the ocean surface that convinced early fishers and explorers that the inscrutable depths contained untold riches. The rude awakening came after industrialized fishing had been going on for a while following World War II; it was perhaps best embodied in the remark of a fisheries expert that "the modal fate of a North Sea fish is to be eaten by a Scandinavian."

Ocean fisheries constitute a paradigm for the "commons"—a common pool, open-access resource available to a large number of harvesters. The catch of each fisher—call him A—imposes a cost, through depletion of the resource. That cost, however, is shared by all fishers, whereas A enjoys the exclusive benefit of his catch and has an incentive to continue even in the face of declining results. This problem is generalized in the classic "Tragedy of the Commons," which we will

SCIENCE IN THE NEWS

Science, Vol. 308, no. 5718, 41–43, 1 April 2005

TAKING THE PULSE OF EARTH'S LIFE-SUPPORT SYSTEMS

Erik Stokstad

The plan was nothing if not ambitious: assess the state of ecosystems across the entire planet, from peat bogs to coral reefs. Rather than solely chart pristine habitats and count species, as many surveys have done, the $20 million Millennium Ecosystem Assessment (MA) put people and their needs front and center. At its core was the question: How well can ecosystems continue to provide the so-called services that people depend on but so often take for granted? These include not just the food and timber already traded on international markets but also assets that are harder to measure in dollar values, such as flood protection and resistance to new infectious diseases.

In another novel approach, the assessment simultaneously examined this issue across a huge range of scales from urban parks to global nutrient cycles. The goal in each case, says director Walter Reid, was to offer policymakers a range of options that might help ecosystems recover or improve their role in providing for human well-being. The participants tried to make clear the tradeoffs involved in managing land; some methods of boosting crop yield, for example, exact a long-term price of degraded soils and incur downstream consequences such as fisheries stunted by fertilizer runoff. What's good for people in one region may cause harm in another place or time.

encounter later. But it is the basis for the phenomena explored in "World Fisheries: The Next 50 Years."

Naturally, the first to experience depletion have been the shallow continental-shelf fisheries that have been historically the most accessible. The groundfish of the New England "banks"—codfish and haddock—were among the first affected, and the plight of those fisheries has presented a chronic political problem for the region. Other in-shore fisheries followed, and the pressure on the industry has gradually moved down in depth and down in the marine food chain, as the illustration in Figure 1 demonstrates dramatically. As fishing effort increased globally, total landings went up, but then they flattened out and started to decline in the 1980s.

What goes on as a fishery comes under increased pressure? Declines in the fish population encourage the development of more efficient technology and encourage larger, longer-range, and higher-capacity vessels. That further intensifies the pressure on the resource, so the process has its own positive feedback. Eventually the failure attracts political and economic attention; the first response, unfortunately, is a government subsidy that only provides incentives for the fishers to stay in business.

Better solutions have been offered, and are at work in different settings. Regulations limiting catch, and/or sparing particular size and age classes to preserve reproductive potential can help. Establishing marine reserves has been successful in a number of places. Many advocates for "market-based" solutions favor tradable permits. That system requires solid estimates of the standing population and of catch needed for sustainability; permits are then issued or auctioned to allow that total "take." In such a system efficient fishers catch their take early and buy permits from the less efficient.

Pauly and his colleagues have made an interesting prediction about market forces. The relationship between energy (fuel costs) and the

> The relationship between energy (fuel costs) and the future of fisheries suggests that if costs rose significantly, the most energy-intensive industrial fisheries, particularly the deep sea-bottom trawlers, might fold. The price of crude oil has nearly doubled since they wrote, and the prediction will bear watching.

future of fisheries suggests that if costs rose significantly, the most energy-intensive industrial fisheries, particularly the deep sea-bottom trawlers, might fold. The price of crude oil has more than doubled since they wrote, and the prediction will bear watching.

Many of these features are among those employed in the thought-provoking scenarios offered in the chapter by Pauly *et al.* These provide an excellent example of the complexity of the problem and the diversity of solutions that might be available. It is clear that fisheries differ widely from one another, and that a one-size solution is unlikely to fit all. Since this essay was written, two national commissions—one organized and funded by the Pew Trust, the other from the U.S. government—have explored solutions for the sustainability of marine resources. The Millennium Ecosystem Assessment also has emerged, with evaluations and recommendations for marine ecosystems.

The special problem of "big fish"—the wide-ranging bluefin tuna, other tuna species, and swordfish—is special in several ways. The incentives for capture are high; in some years an adult bluefin tuna, air-shipped to the Tokyo markets, has brought as much as $30,000. Swordfish—the stars, alas, of *The Perfect Storm*—were for a time a profitable substitute for groundfish for New England fishers. But FDA warnings about mercury as well as warnings about sustainability of the stock have turned off consumers.

In fact the consumer end of the fisheries problem has become a significant project of the environmental movement. Various nongovernmental organizations, university consortia, the Monterey Bay Aquarium, and the David and Lucille Packard Foundation have supported and participated in "Seafood Choice" programs, designed to steer consumers away from fish species with unsustainable populations. Imported swordfish and bluefin tuna have been among the species on the "avoid" list. Public service announcements (for example, "Give Swordfish a Break" and "Pass on Chilean Seabass") have been used, but it will be some time before one can measure their effectiveness.

Aquaculture

On the seafood counter of a local market in Palo Alto, California, there is a striking difference—often as much as a factor of two—between the prices of fish species produced through aquaculture and those caught in the wild by the usual methods. Tilapia, catfish, and rainbow trout are farmed in freshwater ponds; they are among the least expensive, and are recommended in "seafood choice" programs because they are farmed in environmentally sensitive ways. The case for saltwater aquaculture, especially for salmon and shrimp, is not so good. Salmon are raised in pens that float on the sea surface; escapes are frequent, and the result is often that the gene pool of wild relatives is affected. In order to feed these carnivores, large amounts of fishmeal are required—it takes about 4 pounds of food protein to produce each pound of salmon protein. The growing need for fishmeal is one of the factors encouraging "fishing down the food chain" for less desirable species that can be converted to meal. A study published in 2004 in *Science* demonstrated that some farmed salmon contain larger amounts of pesticide residues and other toxicants than wild-caught fish. As for shrimp, large aquaculture operations in Asia and elsewhere have done serious damage to mangrove shorelines, destroying the nurseries for young wild shrimp and invertebrates. The aquaculture pens have to be changed every few years because the water quality becomes degraded, and that means that the assault on the mangrove environment is a continuous process. Although it was once thought that aquaculture would provide a rescue from declining traditional fishery yields, it looks as though only the freshwater sector will make a significant contribution to global food needs, though probably not a large one.

There's a theme here that should be evident by now. The human population is not only growing, but it is clear that per-capita demand for resources and environmental services is growing as well, as societies gradually achieve their developmental aspirations. Look at China: in a 20-year period, meat consumption per capita in that country—mostly pork and chicken—increased ten-fold. That means an increased requirement for grain to feed the livestock, enlarging the human footprint on the land for agriculture. Automobile sales are skyrocketing, and air-quality problems are becoming more severe. Of course, not all Chinese are sharing in an economic growth event that had achieved annual percentage gains of 10 percent or more. But the growth—and the environmental pressure it creates—is impressive to everyone who has seen it.

As we turn from living to nonliving resources, the picture shifts. With respect to the nonliving ones, it is the United States that still emits the most greenhouse gases and still exerts the most demands on the world's energy supplies. The next section will, we hope, focus the reader's attention on what may be required to generate agreement among the industrial nations of the North and the developing nations of the South with respect to the distribution and use of water, oil, toxic emissions, and other commons problems that relate to nonliving resources.

References and Notes

1. G. C. Daily, Ed. *Nature's Services: Societal Dependence on Natural Ecosystems.* Island Press, Washington, 1997.

2. Steadman, D. W. 1995. Prehistoric extinctions of Pacific island birds: Biodiversity meets zooarchaeology. *Science* 267(5201): 1123–1131.

3. G. C. Daily and K. Ellison. *The New Economy of Nature: The Quest to Make Conservation Profitable.* Island Press, Shearwater Books, Washington, 2002.

Human Population
The Next Half Century

JOEL E. COHEN

By 2050, the human population will probably be larger by 2 to 4 billion people, growing more slowly (declining in the more developed regions), more urban, especially in less developed regions, and older than in the 20th century. Two major demographic uncertainties in the next 50 years are international migration and the structure of families. Economies, nonhuman environments, and cultures (including values, religions, and politics) strongly influence demographic changes. Hence, human choices—individual and collective—will have demographic effects, intentional or otherwise.

It is a convenient but potentially dangerous fiction to treat population projections as independent factors in economic, environmental, cultural, and political scenarios, as if population processes were autonomous. Belief in this fiction

This article first appeared in *Science* (14 November 2003: Vol. 302, no. 5648). It has been revised for this edition.

is encouraged by conventional population projections, which ignore food, water, housing, education, health, physical infrastructure, religion, values, institutions, laws, family structure, domestic and international order, and the physical and biological environment. The United Nations Population Division formally recognizes the impact of other biological species on humans explicitly only when quantifying the devastating demographic effects of HIV and AIDS. The absence from population projection algorithms of many influential external variables indicates scientific ignorance of how these variables influence demographic rates rather than a lack of influence (1).

Demographic projections stimulate fears of overpopulation in some people and fears of demographic decline and cultural extinction in others (2). This chapter does not attempt to assess the implications of likely demographic changes for health, nutrition, prosperity, international security, the physical, chemical and biological environment, or human values. Other chapters in this book cover such topics. Our objective here is to review current projections for the next half century to frame later contributions.

Past Population

In about 300 years, Earth's population grew more than 10-fold, from 600 million people in 1700 to 6.3 billion in 2003 (3). These and all demographic statistics are estimates; repeated qualifications of uncertainty will be omitted. It took from the beginning of time until about 1927 to put the first 2 billion people on the planet; less than 50 years to add the next 2 billion people (by 1974); and just 25 years to add the next 2 billion (by 1999). The population doubled in the most recent 40 years. Never before the second half of the 20th century had any person lived through a doubling of global population. Now some have lived through a tripling. The human species lacks any prior experience with such rapid growth and large numbers of its own species.

From 1750 to 1950, Europe and the New World experienced the most rapid population growth of any region, while the populations of most of Asia and Africa grew very slowly. Since 1950, rapid population growth shifted from Western countries to Africa, the Middle East, and Asia.

The most important demographic event in history occurred around 1965–70. The global population growth rate reached its all-time peak of about 2.1 percent per year (pa). It then gradually fell to 1.2 percent pa by 2002 (4). The global **total fertility rate** fell from 5 children per woman per lifetime in 1950–55 to 2.7 children in 2000–05. The absolute annual increase in population peaked around 1990 at 86 million and has fallen to 77 million. In 1960, five countries had total fertility rates at or below the level required to replace the population in the long run. By 2000, there were 64 such countries; together they included about 44 percent of all people (4, 5). Concurrent trends included worldwide efforts to make contraception and reproductive health services available, improvements in the survival of infants and children, widespread economic development and

integration, movements of women into the paid labor market, increases in primary and secondary education for boys and girls, and other cultural changes.

Worldwide urbanization, which has taken place for at least two centuries, accelerated greatly in the 20th century. In 1800, roughly 2 percent of people lived in cities; in 1900, 12 percent; in 2000, more than 47 percent, and nearly 10 percent of those city dwellers lived in cities of 10 million people or larger. Between 1800 and 1900, the number of city dwellers rose more than 11-fold, from 18 million to 200 million; between 1900 and 2000, the number of city dwellers rose another 14-fold or more, from 200 million to 2.9 billion. In 1900, no cities had 10 million people or more. By 1950, one city did: New York. In 2000, 19 cities had 10 million people or more. Of those 19 cities, only four (Tokyo, Osaka, New York, and Los Angeles) were in industrialized countries (6).

Demographic Projections of the Next 50 Years

Projections of future global population prepared by the United Nations Population Division, the World Bank, the United States Census Bureau, and some research institutions assume business as usual (7–9). They include recurrent catastrophes to the extent that such catastrophes are reflected in past trends of **vital rates**, but exclude catastrophes of which there is no prior experience, such as thermonuclear holocaust or abrupt, severe climate change. The following summary relies mainly on the United Nations Population Division's urbanization forecasts (6) and *World Population Prospects: The 2002 Revision* (4).

Estimates of present levels of demographic variables are based on measurements taken in recent years, rather than global current measurements. Using these estimates, the United Nations (UN) prepares several different alternative population projections, including low-, medium-, high-, and constant-fertility scenarios or variants. According to the medium variant, the world's population is expected to grow from 6.3 billion in 2003 to 8.9 billion in 2050. Whereas the first absolute increase by 1 billion people took from the beginning of time until about 1800, the increase by 1 billion people from 6.3 billion to 7.3 billion is projected to require 13 to 14 years. The anticipated increase, by 2050, of 2.6 billion over 2003's population exceeds the total population of the world in 1950, which was 2.5 billion.

Current absolute and relative global population growth rates are far higher than any experienced before World War II. The annual addition of 77 million people poses formidable challenges of food, housing, education, health, employment, political organization, and public order. Virtually all of the increase is and will be in economically less-developed regions.

More than half of the annual increase currently occurs in six countries, from most to least: India, China, Pakistan, Bangladesh, Nigeria, and the United States. Of the total annual increase, the United States accounts for 4 percent.

Were fertility to remain at present levels, the population would grow to 12.8 billion by 2050—more than double its present size. The medium projection of 8.9 billion people in 2050 assumes

that efforts to make means of family planning available to women and couples will continue and will succeed. It also assumes that after 2010, high-risk behaviors related to AIDS will become less frequent and chances of infection among those engaging in high-risk behaviors will decline. The UN's 2002 estimate of 8.9 billion people in 2050 is 0.4 billion lower than that in their 2000 medium variant. About half of the decrease in the projection for 2050 is due to fewer projected births and about half to more projected deaths, notably from AIDS.

Global statistics conceal vastly different stories in different parts of the world. In 2000, about 1.2 billion people lived in economically rich, more-developed regions: Europe, North America, Australia, New Zealand, and Japan. The remaining 4.9 billion lived in economically poor, less-developed regions. The current annual growth rate of global population is 1.22 percent, but wealthier regions' population currently increases 0.25 percent annually while poor regions' population grows 1.46 percent annually—nearly six times faster. The population of the least-developed regions—the 49 countries where the world's poorest 670 million people lived in 2000—increases annually by 2.41 percent. By 2050, the annual growth rate of global population is projected to be 0.33 percent. The poor countries' population will still be increasing 0.4 percent annually, whereas the population of the rich countries will have been declining for 20 years and will then be falling at −0.14 percent pa.

Thirty of the more developed countries are expected to have lower populations in 2050 than today, including Japan (expected to be 14 percent smaller), Italy (22 percent smaller), and the Russian Federation (29 percent smaller). By contrast, the population of today's poor countries is projected to rise to 7.7 billion in 2050 from 4.9 billion in 2000. Fertility in the less-developed regions is expected to fall to replacement level in 2030–35, but to remain above 2 children per woman by 2050 because some of the least-developed countries will still have total fertility rates well above replacement level. The population of these high-fertility poor countries will be an increasing proportion of the population of the less developed regions.

The world's average population density of 45 people/km^2 in 2000 is projected to rise to 66 people/km^2 by 2050. Globally, perhaps 10 percent of land is arable, so population densities per unit of arable land are roughly 10 times higher. In the rich countries, the population density was 23 people/km^2 in 2000—half the global average—and was projected not to change at all by 2050. In the poor countries, the population density was 59 people/km^2 in 2000 and was projected to rise to 93 people/km^2 in 2050. For comparison, the population density of Liechtenstein was 204 people/km^2 in 2000 and that of the United States was 30. A population density of 93 people/km^2 over the entire developing world will pose unprecedented problems of land use and preservation.

According to these projections, the ratio of population density in the poor countries to that in the rich countries is projected to rise from 2.6 in 2000 to 4.0 in 2050. Over the same interval, while the population density of Europe is projected to drop from 32 to 27 people/km^2, that of Africa is projected to rise from 26 to 60 people/km^2. The ratio of population density in Africa to that in Europe is projected to rise from 0.8 in 2000 to 2.2 in 2050. It seems plausible to anticipate increasing human effects on the natural environment in Africa and increasing pressure of migrants from Africa to Europe.

The difference in the population growth rate between rich and poor countries affects both population size and age structure. If a population grows slowly, the number of births each year nearly balances the number of deaths. Because most deaths occur at older ages, the numbers of individuals in different age groups are roughly equal up to older ages. The so-called population pyramid of a slowly growing population resembles a column (Figure 1, middle row left) (10). If a population grows rapidly, each birth cohort is larger than its predecessor and the population pyramid is triangular (Figure 1, middle row right).

The projected difference in age structures between the European Union versus North Africa and western Asia (Figure 1, bottom) has obvious implications for the supplies of military personnel and ratios of elderly to middle-aged.

Inequality in the face of death between rich and poor will decrease but remain large if survival improves everywhere as anticipated in the coming half century. Global life expectancy in 2000–05 is estimated at 65 years; in 2045–50, it is projected at 74 years. Over the same interval, life expectancy in the rich countries is expected to rise from 76 years to 82 years and in the poor countries from 63 years to 73 years. The average infant born in a poor country had a chance of dying before age 1 that was 8.1 times higher than that in a rich country in 2000–05; the same ratio is projected to be 5.2 in 2045–50.

Despite higher death rates, poor countries' populations grow faster than those of rich because birth rates in poor countries are much higher. At current birth rates, during her lifetime, the average woman in the poor countries bears nearly twice as many children (2.9) as in the rich countries (1.6). By 2050, according to the medium variant, the total fertility rate in today's poor countries will drop to 2.0. The total fertility rate in today's more developed countries is projected to rise to almost 1.9 children per woman, as timing effects (beginning childbearing at later ages) that currently depress the total fertility rate cease to operate.

Urbanization will continue to be an important trend. In the coming decade, more than half of all people will live in cities for the first time in human history. Almost all population growth in the next half century will occur in cities in poor countries while the world's rural population will remain flat—near 3 billion people. The United Nations Population Division projects urban population only as far as 2030 (6). Its figures on urbanization disguise major ambiguities and variations among countries, primarily because of differences in definitions of the terms *cities* and *urban*. Nevertheless, the trend toward urbanization is clear. Of the projected 2.2 billion increase in population from 2000 to 2030, 2.1 billion will

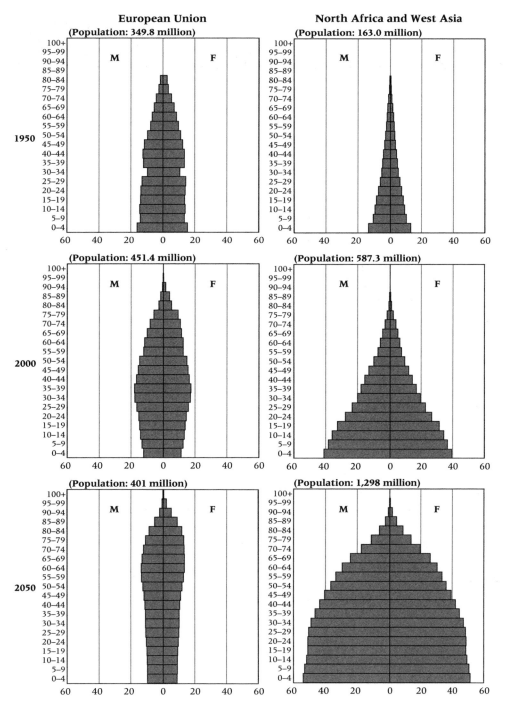

FIGURE 1.   Population size and age distribution for 1950, 2000, and 2050 in an anticipated
enlarged European Union of 25 countries and in 25 countries of North Africa and West Asia
between India's western border and the Atlantic Ocean. The figure excludes countries of
central Asia that were part of the former Soviet Union, those of Muslim black Africa, and
Israel (10). Horizontal scale gives million persons separately by sex; vertical scale gives
age groups in increments of 5 years.

Science, Vol. 309, no. 5731, 102, 1 July 2005

WILL MALTHUS CONTINUE TO BE WRONG?

Erik Stokstad

In 1798, a 32-year-old curate at a small parish church in Albury, England, published a sobering pamphlet entitled *An Essay on the Principle of Population*. As a grim rebuttal of the utopian philosophers of his day, Thomas Malthus argued that human populations will always tend to grow and, eventually, they will always be checked—either by foresight, such as birth control, or as a result of famine, war, or disease. Those speculations have inspired many a dire warning from environmentalists.

Since Malthus's time, world population has risen sixfold to more than 6 billion. Yet, happily, apocalyptic collapses have mostly been prevented by the advent of cheap energy, the rise of science and technology, and the green revolution. Most demographers predict that by 2100, global population will level off at about 10 billion.

Out of balance. Sustaining a growing world population is threatened by inefficient consumption of resources—and by poverty.

CREDIT: NELSON SYOZI/ISTOCKPHOTO.COM

The urgent question is whether current standards of living can be sustained while improving the plight of those in need. Consumption of resources—not just food but also water, fossil fuels, timber, and other essentials—has grown enormously in the developed world. In addition, humans have compounded the direct threats to those resources in many ways, including by changing climate, polluting land and water, and spreading invasive species.

How can humans live sustainably on the planet and do so in a way that manages to preserve some biodiversity? Tackling that question involves a broad range of research for natural and social scientists. It's abundantly clear, for example, that humans are degrading many ecosystems and hindering their ability to provide clean water and other "goods and services." But exactly how bad is the situation? Researchers need better information on the status and trends of wetlands, forests, and other areas. To set priorities, they'd also like a better understanding of what makes ecosystems more resistant or vulnerable and whether stressed ecosystems, such as marine fisheries, have a threshold at which they won't recover.

be in urban areas, and all but 0.1 billion of that urban increase will be in developing countries. The annual rate of increase of urban population over the next 30 years, 1.8 percent, is nearly twice the projected annual rate of increase of global population during that period.

The urban population of developing regions will grow rapidly as people migrate from rural to existing urban areas and transform rural settlements into cities. The rural population of the rich countries peaked around 1950 and has slowly declined since then. The rural population of the presently poor countries is expected to peak around 2025 and then gradually decline. Urbanization of the rich countries will continue, with city dwellers rising from 75 percent of people in 2000 to 83 percent in 2030. Over the same period, urbanization of the poor countries will rise from 40 percent to 56 percent, similar to the level of urbanization in the rich countries in 1950.

The coming half century will also see a dramatic aging of the population, which means that a higher proportion of the population will be in elderly age groups. The proportion of children aged 4 years and under peaked in 1955 at 14.5 percent and gradually declined to 10.2 percent in 2000. By contrast, the proportion of people aged 60 years and older gradually increased from a low of 8.1 percent in 1960 to 10.0 percent in 2000. Each group constitutes about 10 percent of humanity today. The 20th century will probably be the last in which younger people outnumbered older ones. Children aged 0 to 4 are projected to decline to 6.6 percent of global population by 2050, whereas people aged 60 years and older are projected to more than double, to 21.4 percent. By 2050, there will be 3.2 people aged 60 years or older for every child 4 years old or younger.

This reversal in the numerical dominances of old and young reflects improved survival and reduced fertility. Improved survival raised the global average length of life from perhaps 30 years at the beginning of the 20th century to 65 years at the beginning of the 21st. Reduced fertility rates added smaller cohorts to the younger age groups.

Because the populations of the poor countries have been growing more rapidly than those of the rich, they have a much higher fraction of people under the age of 15 years (33 percent versus 18 percent in 2000). By 2050, in the medium variant, these fractions will drop to 21 percent and 16 percent in poor and rich countries, respectively. The global fraction of the elderly population (aged 65 years or more) will rise from 7 percent in 2000 to 16 percent by 2050. Over the

same period, the elderly fraction will rise from 5 to 14 percent in the presently poor countries and from 14 to 26 percent in the rich countries.

Though the fraction of children in the population will decrease by more in the poor countries than in the rich, the fraction of elderly will increase by more in the rich countries than in the poor. Both shifts will have consequences for spending on the young and the old. Slowly growing populations have a higher **elderly dependency ratio** (the ratio of the number of people aged 65

KEY TERM

The **elderly dependency ratio** is the ratio of people 65 and over to the population aged 15–64. A high ratio is interpreted to mean that those of working age face a greater burden in supporting the aging population, but that interpretation depends heavily on the age of retirement from work and the health of the elderly.

and older to the number aged 15 to 64), while rapidly growing populations have a higher youth dependency ratio (the ratio of the number of people aged 0 to 14 to the number aged 15 to 64).

The elderly dependency ratio rose from 1950 to 2000 rapidly in the more developed countries, slightly less rapidly in the United States, and still less rapidly in the world as a whole. The ratio rose only slightly in the less-developed countries, and hardly at all in the least-developed countries. After 2010, in the more developed countries, the United States, and the less-developed countries, the elderly dependency ratio will increase much faster than in the past; this acceleration will be greater in the more developed countries and the United States. The least-developed countries will experience a slow increase in the elderly dependency ratio after 2020 and, by 2050, will be approaching the elderly dependency ratio found in the more-developed countries in 1950.

Demographic Uncertainties: Migration and the Family

According to the United Nations Population Division, "International migration is the component of population dynamics most difficult to project reliably. This occurs in part because the data available on past trends are sparse and partial, and in part because the movement of people across international boundaries, which is a response to rapidly changing economic, geopolitical or security factors, is subject to a great deal of volatility" (11). The UN's 2002 medium variant posits migration from less- to more-developed regions of 2.6 million people annually during 1995–2000, declining to nearly 2.0 million by 2025–30, and remaining constant at that level until 2050. The United States is anticipated to increase by 1.1 million of these 2 million migrants annually, more than five times the number expected to be added to the next largest recipient, Germany (at 211,000 migrants annually). The major sending countries are expected to be China, Mexico, India, the Philippines, and Indonesia, in decreasing order.

International migration is likely to remain important for specific countries, including the United States. In the mid-1990s, about 125 million people (2 percent of world population) resided outside of their country of birth or citizenship. In 1990, only 11 countries in the world had more than 2 million migrants; collectively they had almost 70 million migrants. The largest numbers of migrants were in the United States (19.6 million), India (8.7 million), Pakistan (7.3 million), France (5.9 million), and Germany (5.0 million). The countries with the highest percentage of international migrants in the total population were countries with relatively small populations. In the United Arab Emirates, Andorra, Kuwait, Monaco, and Qatar, 64 to 90 percent of the population were immigrants.

Predicting international migration is difficult. Predicting change in family structure is even more difficult. Goldscheider (12) suggested that the fall in fertility during the demographic transition weakened the ties between men and women based on parenthood and that the rise in divorce and cohabitation is weakening the ties between fathers and children. Nonmarital births increased as a percentage of all births in the United States from 5.3 percent in 1960 to 33.0 percent in 1999. In 1999, the United States had 1.3 million births to unmarried women (13). In 1998, Iceland, Norway, Sweden, Denmark, France, the United Kingdom, and Finland all had higher proportions of nonmarital births than the United States. By contrast, in Germany, Italy, Greece, and Japan, less than 15 percent of births were nonmarital (13). Among United States women aged 15 to 29 years at first birth, when that first birth was conceived before marriage, the fraction who married before the birth fell from 60 percent in 1960–64 to 23 percent in 1990–94 (14). By 1994, about 40 percent of children in the United States did not live with their biological father (12).

In the United States, the number of widowed males aged 55 to 64 per thousand married persons fell from 149 in 1900 to 35 in 2000, whereas the number of divorced men aged 55 to 64 per thousand married persons rose from 7 to 129. Divorced men became more frequent than widowed men between 1970 and 1980. Divorced women became more frequent than widowed women between 1990 and 2000. By 2000, the number of divorced and widowed persons aged 55 to 64 per thousand married persons was 164 men and 426 women (2.6 such women for each such man) (15). Remarriages and stepfamilies are becoming increasingly common.

Three factors set the stage for further major changes in families: fertility falling to very low levels; increasing longevity; and changing mores of marriage, cohabitation, and divorce. In a population with one child per family, no children have siblings. In the next generation, the children of those children have no cousins, aunts, or uncles. If adults live 80 years and bear children between age 20 and 30 on average, then the parents will have decades of life after their children have reached adulthood and their children will have decades of life with elderly parents. The full effects on marriage, child bearing, and child rearing of greater equality between the sexes in education; earnings; and social, legal, and political rights have yet to be felt or understood.

References and Notes

1. J. E. Cohen, *How Many People Can the Earth Support?* (W. W. Norton, New York, 1995).

2. L. Shriver, *Popul. Dev. Rev.* 29 (no. 2), 153 (2003).

3. United States Census Bureau, *Historical Estimates of World Population* (online). Available at www.census.gov/ipc/www/worldhis.html (cited 21 June 2003).

4. United Nations Population Division, *World Population Prospects: The 2002 Revision, Highlights* (online database). ESA/P/WP.180, revised 26 February 2003, p. vi. Available at: http://esa.un.org/unpp/ (consulted 1–30 June 2003).

5. United Nations Population Division, *Partnership and Reproductive Behaviour in Low-Fertility Countries*, ESA/P/WP.177, revised May 2003. Available at www.un.org/esa/population/publications/reprobehavior/partrepro.pdf (cited 29 June 2003).

6. United Nations Population Division, *World Urbanization Prospects: The 2001 Revision.* ESA/P/WP.173. (United Nations, New York, 2002).

7. J. Bongaarts, R. A. Bulatao, Eds. *Beyond Six Billion: Forecasting the World's Population* (National Academy Press, Washington, DC, 2002).

8. J. E. Cohen, in *Seismic Shifts: The Economic Impact of Demographic Change*, J. S. Little, R. K. Triest, Eds., Federal Reserve Bank of Boston, conference series no. 46, 11–13 June 2001 (Federal Reserve Bank of Boston, Boston, MA, 2001), pp. 83–113.

9. J. E. Cohen, in *What the Future Holds: Insights from Social Science*, R. N. Cooper, R. Layard, Eds. (MIT Press, Cambridge, MA, 2002), pp. 29–75.

10. P. Demeny, *Popul. Dev. Rev.* 29 (no. 1), 1 (2003).

11. United Nations Population Division, *World Population Prospects: The 1998 Revision: Volume III: Analytical Report.* ESA/P/WP.156, revised 18 November 1999 (United Nations, New York, 1999).

12. F. K. Goldscheider, *Futurist* 32, 527 (2000).

13. S. J. Ventura, C. A. Bachrach, *National Vital Statistics Reports* 48 (no. 16, revised), 18 October 2000. Available at www.cdc.gov/nchs/data/nvsr/nvsr48/nvs48_16.pdf (cited 25 June 2003).

14. A. Bachu, *Current Population Reports; P23–197* (U.S. Census Bureau, Washington, DC, 1999).

15. P. Uhlenberg, in *United Nations, Department for Economic and Social Information and Policy Analysis, 1994. Ageing and the Family.* Proceedings of the United Nations International Conference on Ageing Populations in the Context of the Family, ST/ESA/SER.R/124 (United Nations, New York, 1994), pp. 121–127.

16. I acknowledge with thanks the support of U.S. National Science Foundation grant DEB 9981552, the assistance of P. K. Rogerson, and the hospitality of Mr. and Mrs. W. T. Golden during this work.

Prospects for Biodiversity

MARTIN JENKINS

Assuming no radical transformation in human behavior, we can expect important changes in biodiversity and **ecosystem services** by 2050. A considerable number of species extinctions will have taken place. Existing large blocks of tropical forest will be much reduced and fragmented, but temperate forests and some tropical forests will be stable or increasing in area, although the latter will be biotically impoverished. Marine ecosystems will be very different from today's, with few large marine predators, and freshwater **biodiversity** will be severely reduced almost everywhere. These changes will not, in themselves, threaten the survival of humans as a species.

What will be the state of the world's biodiversity in 2050, and what goods and services can we hope to derive from it? First, some assump-

tions: that the United Nations median population estimate for 2050 holds, so that Earth will have roughly nine billion people—just under half again as many as are currently alive (1, 2); that the Intergovernmental Panel on Climate Change scenarios provide a good indication of global

This article first appeared in *Science* (14 November 2003: Vol. 302, no. 5648). It has been revised and updated for this edition.

KEY TERM

Ecosystem services: These include provisioning services of such goods as food and water; regulating services such as flood and disease control; cultural services such as spiritual, recreational, and cultural benefits; and supporting services such as nutrient cycling that maintain the conditions for life on Earth.

average surface temperatures and atmospheric CO_2 concentrations at that time, with the former 2.0°C to 5.4°C and the latter ~100 to 200 parts per million higher than today (3); and, perhaps most important, although most nebulous, that humanity as a whole has not determined on a radically new way of conducting its affairs. Here, then, is a plausible future.

In this future, the factors that are most directly implicated in changes in biodiversity—habitat conversion, exploitation of wild resources, and the impact of introduced species (4)—will continue to exert major influences, although their relative importance will vary regionally and across biomes. In combination, they will ensure continuing global biodiversity loss, as expressed through declines in populations of wild species and reduction in area of wild habitats.

Extinction Rates

To start, as it were, at the end: with extinction, which is perhaps the most tangible measure of biodiversity loss. Together, the uncertainties that still surround our knowledge of tropical biotas (which include the great majority of extant species); the difficulty of recording extinctions; and our ability, when we put our minds to it, to bring species back from the brink make it extremely difficult to assess current global extinction rates, let alone estimate future ones. However, an assessment of extinction risk in birds carried out by BirdLife International—using the criteria of IUCN, the World Conservation Union's Red List of Threatened Species—has concluded (with many caveats) that perhaps 350 species (3.5 percent of the world's current avifauna) might be expected to become extinct between now and 2050 (5). Indications are that some other groups—mammals and freshwater fishes, for example—have a higher proportion of species at risk of extinction, although data for these are less complete (4).

Just as it is hard to estimate future extinction rates, so is it difficult to extrapolate forward from current rates of habitat alteration, even where these are known (6). However, some general patterns are clear. With the harvest of marine resources now at or past its peak (7), terrestrial ecosystems will bear most of the burden of having to feed, clothe, and house the expanded human population. This extra burden will fall most heavily on developing countries in the tropics, where the great majority of the world's terrestrial biological diversity is found.

The Land

Most increased agricultural production is expected to be derived from intensification. However, the Food and Agriculture Organization (FAO) of the United Nations notes that, on the basis of reasonably optimistic assumptions about increasing productivity, at least an extra 120 million hectares of agricultural land will still be needed in developing countries by 2030 (8). In a less than wholly efficient world, the amount converted to agricultural production will be much more. Historic precedent and present land availability indicate that almost all new conversion will be in South America and sub-Saharan Africa. More than half the unused suitable cropland is found in just seven countries in these regions: Angola, Argentina, Bolivia, Brazil, Colombia, Democratic Republic of Congo, and Sudan (8).

SCIENCE IN THE NEWS

Science, Vol. 309, no. 5734, 546–547, 22 July 2005

GLOBAL ANALYSES REVEAL MAMMALS FACING RISK OF EXTINCTION
Erik Stokstad

Two new studies are helping conservation biologists think big—in the case of one of the studies, as big as one-tenth of the continents.

Conservationists typically set goals and priorities for relatively small regions. Although some have come up with priorities for the planet, these have often been wish lists rather than objectives drawn from rigorous analyses. Now a team of researchers, led by mammalogist Gerardo Ceballos of the National Autonomous University of Mexico, has conducted the first global analysis of the conservation status of all known land mammals. They report that 25 percent of known mammal species are at risk of extinction. In order to decrease the risk to mammals worldwide, about 11 percent of Earth's land should be managed for conservation, the analysis finds.

Big risk. Large size significantly ups the odds of extinction for mammals such as elephants and pandas.
CREDIT: LEFT: CORBIS, RIGHT: WANG SANJUN/ISTOCKPHOTO.COM

This is the first time such a global conservation estimate has been calculated for mammals, and although experts are not surprised by these results, they praise the study for its comprehensiveness and detail. "This sets a new standard for global priority-setting analyses," says Peter Kareiva, lead scientist for The Nature Conservancy.

A second conservation study finds that large mammals may be more threatened than their smaller relatives. A team led by Georgina Mace of the Zoological Society of London and Andy Purvis of the Imperial College London reports that adult mammals that weigh more than 3 kilograms tend to have biological traits that hike their risk of extinction. "Both of these papers provide us with finer and more detailed insights into threat patterns and processes," says Thomas Brooks of Conservation International in Washington, D.C.

Five of these are among the 25 most biodiverse countries; the exceptions (Angola and Sudan) are both also highly biodiverse (9, 10). Large-scale conversion of uncultivated to cultivated land will continue in most or all of these countries, with a disproportionately high impact on global biodiversity.

Much conversion here and elsewhere will be of land that is currently tropical forest. Fragmentation and loss of such forests will thus continue, albeit overall possibly at a slower rate than at present. The great, largely contiguous forest blocks of Amazonia and the Zaire basin will by

2050 be a thing of the past, with unknown (and hotly debated) effects on regional weather patterns and global climate. Deforestation pressure will remain high in the immediate future in a number of other tropical developing countries. Among these countries are those such as Indonesia, Madagascar, and the Philippines, which hold many forest-dependent endemic species, often with small ranges (11, 12). Forest loss here will also have a particularly high impact on biodiversity.

There will, however, still be considerable for-

KEY TERM

Endemic species are those found in only one geographic area. Species endemism tends to be concentrated in just a few, narrow geographic regions, sometimes known as hotspots; loss of a hotspot causes higher loss in biodiversity because these species are not found elsewhere.

est cover in the tropics, much of it in inaccessible or steeply sloping sites unsuitable for clearance and some in protected areas. Even outside such areas, forest cover will be increasing in some regions—paralleling the current situation in Northern Hemisphere temperate forests (13)—because growing urbanization will lead to the abandonment of marginally productive lands (1), allowing them to revert to a more natural state. However, uncontrolled and frequent fires will mean that abandoned lands in many areas will remain relatively degraded. In addition, almost all wild lands in the tropics will be impoverished in numbers and diversity of larger animal species, thanks to persistent overexploitation of wild resources such as bushmeat.

Although there have been some local successes, the goal of large-scale sustainable harvest of these resources has so far been elusive and will remain so (14). This means that populations of

many species will survive largely or exclusively in heavily managed protected areas.

Although tropical developing countries will continue to suffer quite possibly accelerating biodiversity loss, much less change can be expected in developed temperate countries. Temperate forest cover will continue to increase, or at least stabilize. Many forest species will thrive, although with changes in distribution and relative abundance as a result of climate change. The recent declines in many wild species that are primarily associated with agricultural land (15) may or may not continue. Much will depend on whether the current consumer-led drive to "greener" forms of agriculture has a major long-term impact.

Aquatic Ecosystems

Our most direct and pervasive impact on marine ecosystems and marine biodiversity is through fishing. If present trends [reviewed in detail in (7)] continue, the world's marine ecosystems in 2050 will look very different from today's. Large species, particularly top predators, will by and large be extremely scarce, and some will have disappeared entirely, giving the lie to the old assertion that marine organisms are peculiarly resistant to extinction.

Marine ecosystems, particularly coastal ones, will also continue to contend with a wide range of other pressures, including siltation and eutrophication from land runoff, coastal development, conversion for aquaculture, and effects of climate change (9). Areas of anoxia will increase; most coral reefs will be heavily degraded, but some adaptable species may benefit from warming and may even have started to expand in range.

Available information suggests that freshwater biodiversity has declined as a whole faster than either terrestrial or marine biodiversity over the past 30 years (Figure 2) (16). The increasing demands that will be placed on freshwater resources in most parts of the world mean that this uneven loss of biodiversity will continue (17). Pollution, siltation, canalization, water

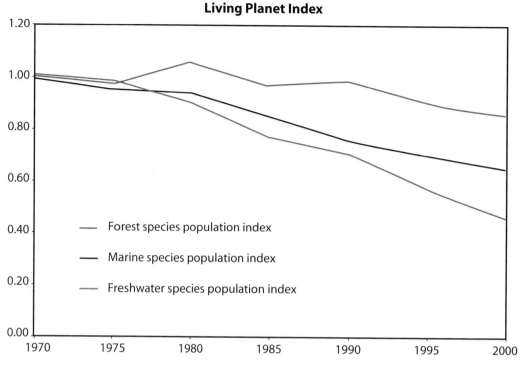

FIGURE 2. Species population indices from 1970 to 2000 for forest, marine, and freshwater ecosystems, as included in the 2002 WWF Living Planet Index. Data for 1996 to 2000 are drawn from small samples (16).

abstraction, dam construction, overfishing, and introduced species will all play a part in the decline, although their individual effects will vary regionally. The greatest impact will be on biodiversity in fresh waters in densely populated parts of the tropics, particularly South and Southeast Asia, and in dryland areas, although large-scale hydroengineering projects proposed elsewhere could also have catastrophic effects (18). Although water quality may stabilize or improve in many inland water systems in developed countries, other factors—such as introduced species—will continue to have an adverse impact on biodiversity in most areas.

How Much Does It Matter?

In assessing the importance of environmental change, we must distinguish between wholesale degradation—such as reduction of a productive, forested slope to bedrock—and reduction in bio-

diversity per se through the loss of particular populations or species of wild organisms or the replacement of diverse, species-rich systems with less diverse, often intensively managed systems of nonnative species. The former can, of course, have devastating direct consequences for human well-being.

It is much more difficult to determine the effects of the latter. In truth, ecologists and conservationists have struggled to demonstrate the increased material benefits to humans of "intact" wild systems over largely anthropogenic ones. In terms of the most direct benefits, the reverse is indeed obviously the case; this is the logic that has driven us to convert some 1.5 billion hectares of land area to highly productive, managed, and generally low-diversity systems under agriculture.

Even with regard to indirect ecological services such as carbon sequestration, regulation of water flow, and soil retention, it seems that there are few cases in which these services cannot adequately be provided by managed, generally low-diversity, systems. Where increased benefits of natural systems have been shown, they are usually marginal and local (19).

Nowhere is this more starkly revealed than in the extinction of species. There is growing consensus that from around 40,000 to 50,000 years ago onward (20), humans have been directly or indirectly responsible for the extinction in many

KEY TERM

Anthropogenic systems: Ecosystems are functional units that result from the interactions of abiotic, biotic, and cultural (or anthropogenic) components. Ecosystems are a combination of interacting, interrelated parts that form a unitary whole, but Earth's ecosystems are increasingly dominated by their cultural component.

parts of the world of all or most of the larger terrestrial animal species. Although these species were only a small proportion of the total number of species present, they undoubtedly exerted a major ecological influence (21, 22). This means that the "natural" systems we currently think of in these parts of the world (North and South America, Australasia, and virtually all oceanic islands) are nothing of the sort, and yet they still function—at least according to our perceptions and over the time scales we are currently capable of measuring. In one well-documented case, New Zealand, a flightless avifauna of at least 38 species has been reduced in a few centuries to 9, most of which are endangered. Here, as David Steadman recently put it, "much of the biodiversity crisis is over. People won: Native plants and animals lost" (23). Yet, from a functional perspective, New Zealand shows few signs overall of suffering ter-

minal crisis. There is currently little evidence to dissuade us from the view that what applies for New Zealand today could equally hold, more or less, for the world as a whole tomorrow.

This does not mean, of course, that we can continue to manipulate or abuse the biosphere indefinitely. At some point, some threshold may be crossed, with unforeseeable but probably catastrophic consequences for humans. However, it seems more likely that these consequences would be brought about by other factors, such as abrupt climate shifts (24), albeit ones in which ecosystem changes may have played a part.

References and Notes

1. J. Cohen, *Science* 302, 1172 (2003).
2. *World Population Prospects: The 2002 Revision* (United Nations, ESA/P/WP. 180).
3. Intergovernmental Panel on Climate Change (IPCC), *Third Assessment Report* (IPCC, Geneva, 2001).
4. See summary data on threatened species at www.redlist.org.
5. BirdLife International, *Threatened Birds of the World* (Lynx Edicions and BirdLife International, Barcelona and Cambridge, UK, 2000).
6. M. Jenkins *et al.*, *Conserv. Biol.* 17, 20 (2003).
7. D. Pauly *et al.*, *Science* 302, 1359 (2003).
8. J. Bruinsma, Ed., *World Agriculture: Towards 2015/2030, an FAO Perspective* (Earthscan, London, 2003).
9. B. Groombridge, M. D. Jenkins, *World Atlas of Biodiversity* (Univ. California Press, Berkeley, 2002).
10. J. O. Caldecott *et al.*, *Biodivers. Cons.* 5, 699 (1996).
11. A. Balmford, A. Long, *Nature* 372, 623 (1994).
12. P. Jepson *et al.*, *Science* 292, 859 (2001).
13. Food and Agriculture Organization of the United Nations, *Global Forest Resources Assessment 2000*, FAO Forestry Paper 140.
14. N. Leader-Williams, in *Guidance for CITES Scientific Authorities*, A. Rosser, M. Haywood, compilers. Occasional Paper of the IUCN Species Survival Commission No. 27 (2002).
15. See, for example, the United Kingdom headline indicators of sustainable development at www.sustainabledevelopment.gov.uk/indicators/headline/index.htm.
16. J. Loh *et al.*, *Living Planet Report 2002* (WWF, World Wide Fund for Nature, Gland, Switzerland, 2002).
17. *The United Nations World Water Development Report* (UNESCO & Berghahn Books, Paris and Oxford, 2003).
18. World Conservation Monitoring Centre (WCMC),

Freshwater Biodiversity: A Preliminary Global Assessment (WCMC, World Conservation Press, Cambridge, 1998).

19. A. Balmford *et al.*, *Science* 297, 950 (2002).

20. R. G. Roberts *et al.*, *Science* 292, 1888 (2001).

21. P. S. Martin, R. G. Klein, Eds., *Quaternary Extinctions* (University of Arizona Press, Tucson, AZ, 1984).

22. W. Schüle, in *Tropical Forests in Transition*, J. G. Goldhammer, Ed. (Birkhaüser Verlag, Basel, 1992), pp. 45–76.

23. D. W. Steadman, *Science* 298, 2136 (2002).

24. R. B. Alley *et al.*, *Science* 299, 2005 (2003).

25. I thank two anonymous reviewers and many friends, in particular N. Ash, H. Coulby, J. Gavrilovic, T. Hufton, V. Kapos, J. Loh, K. Mackinnon, E. McManus, A. Rosser, A. Stattersfield, and K. Teleki, for discussion and comments.

Web Resources

www.sciencemag.org/cgi/content/full/302/5648/1175/DC1

World Fisheries

The Next 50 Years

DANIEL PAULY, JACKIE ALDER, ELENA BENNETT, VILLY CHRISTENSEN, PETER TYEDMERS, and REG WATSON

Formal analyses of long-term global marine fisheries prospects have yet to be performed because fisheries research focuses on local, species-specific management issues. Extrapolation of present trends implies expansion of **bottom fisheries** into deeper waters, serious impact on biodiversity, and declining global catches—the last possibly aggravated by fuel cost increases. Examination of four scenarios, covering various societal development choices, suggests that the negative trends now besetting fisheries can be turned around, and their supporting ecosystems rebuilt, at least partly.

Fisheries are commonly perceived as local affairs requiring, in terms of scientific inputs, annual reassessments of species-specific catch quotas. Most fisheries scientists are employed by

This article first appeared in *Science* (21 November 2003: Vol. 302, no. 5649) as "The Future for Fisheries." It has been revised and updated for this edition.

regulatory agencies to generate these quotas—which ideally should make fisheries sustainable and profitable, contributors to employment and, through international trade, contributors to global food security.

This perception of fisheries as local and species-specific, managed to directly benefit the

KEY TERM

Bottom fisheries: These are generally for groundfish species such as rockfish, cod, and sole, which are caught by bottom trawling, the fishing practice of dragging large nets weighted with chains, rollers, or rock-hopper gear across the seafloor. Trawling is the fishing practice most destructive of marine habitats.

SCIENCE IN THE NEWS

Science, Vol. 309, no. 5742, 1806, 16 September 2005

SCIENCE IN IRAN: THE STURGEON'S LAST STAND
Richard Stone

RASHT, IRAN—In a cavernous hall packed with naval-gray steel tanks, a precious commodity is being enriched and multiplied. No, this is not a hitherto undeclared uranium facility in Iran's nuclear program: It's a breeding facility for Caspian sturgeons. Each tank is filled with fish of various ages, from fingerlings, a few centimeters long with crocodilian snouts, to meter-long juveniles. Here at the International Sturgeon Research Institute (ISRI) in the northern town of Rasht, scientists are refining techniques for rearing fingerlings that may give the ancient but threatened species a better shot at surviving in the open sea. "If something happened in the Caspian and a wild population was lost, we could reconstitute it," says ISRI director Mohammad Pourkazemi.

Caspian Sturgeon. Beluga, Russian, and stellate sturgeon are caught by Kazakh fishermen in the Ural River, possibly the last place where belugas reproduce in the wild.

CREDIT: SHANNON CROWNOVER, SEAWEB.
COURTESY OF CAVIAR EMPTOR

ISRI may get a chance to test that claim: Deteriorating spawning grounds and unbridled poaching have reduced sturgeon stocks to a shadow of what they were a generation ago. With disaster looming, the two biggest fishing nations—Iran and Russia—are sparring over how many sturgeon are left and how to divvy up a declining catch. Amid the bickering, a new survey suggests that the sturgeon's free fall is continuing.

For most of the 20th century, Iran and the Soviet Union ran tight ships, at least on regulating sturgeon fisheries. The situation unraveled in 1991 when four Caspian states—Azerbaijan, Kazakhstan, Russia, and Turkmenistan—emerged from the Soviet collapse. Weak law enforcement and poverty along the Volga and Ural rivers, the northern spawning areas, have enabled poachers to take up to 10 times the legal catch. Despite the release of tens of millions of fingerlings each year, Caspian nations in 2004 caught only 760 tons of sturgeon, the smallest figure in a century, down from 26,600 tons in 1985.

Four Caspian sturgeon varieties—Russian, Persian, beluga, and stellate, or sevruga—supply 90 percent of the world's caviar. A fifth, the ship sturgeon, is so scarce that exporting its meat or caviar has been banned since 2002. Almost half of this year's caviar quota—51 of 105 tons—is Persian sturgeon, which mostly keeps to Iranian waters. Iranian officials attribute its relative robustness to government control of the caviar trade and zero tolerance for poaching.

But the Persian's rise has come at the expense of its kin, throwing the ecosystem off kilter, asserts marine ecologist Arash Javanshir of the University of Tehran. Besides fishing restrictions, he says, what's needed is a restoration program by the Caspian states that targets all sturgeon and their spawning grounds and many other organisms as well.

fishers themselves, is conducive to neither global predictions nor the collaborative development of long-term scenarios. Indeed, recent accounts of this type, except those of the United Nations Food and Agriculture Organization (FAO) (1), tend to be self-conscious and layered in irony (2–5), perhaps an appropriate response to 19th-century notions of inexhaustibility.

The past decade established that fisheries must be viewed as components of a global enterprise, on its way to undermining its supporting ecosystems (6–11). These developments occur against a backdrop of fishing industry lobbyists arguing that governments drop troublesome regulations and economists assuming that free markets generate inexhaustibility. The aquaculture sector offers to feed the world with farmed fish, while building more coastal feedlots wherein carnivores such as salmon and tuna are fed with other fish (12), the aquatic equivalent of robbing Peter to pay Paul.

The time has come to look at the future of fisheries through (i) identification and extrapolation of fundamental trends and (ii) development and exploration (with or without computer simulation) of possible futures.

The fisheries research community relied, for broad-based analyses, on a data set now shown to be severely biased (10). First-order correction suggests that rather than increasing, as previously reported, global fisheries landings instead are declining by about 500,000 metric tons per year from a peak of 80 to 85 million tons in the late 1980s. Because overfishing and habitat degradation are likely to continue, extrapolation may be considered (see below). This correction, however, does not consider discarded **bycatch** to be only one component of the illegal, unreported, or unrecorded (IUU) catches that recently became part of the international fisheries research agenda (13–17).

The geographic and depth expansion of fisheries is easier to extrapolate (Figure 3). Over the past 50 years, fisheries targeting **benthic** and bentho-**pelagic** organisms have covered the shelves surrounding continents and islands down to 200 m, with increasing inroads below

KEY TERMS

Bycatch refers to species caught in a fishery that is targeting another species. Bycatch is often discarded and is usually dead or dying by the time it is returned to the sea.

The benthos is an aggregation of organisms living on or at the bottom of a body of water. **Benthic** organisms include infauna—the animals, and bacteria that live in the sediment; epifauna—organisms that are attached to or live on the hard-bottom or substrate (for example, rocks or debris); and demersals—the bottom-feeding or bottom-dwelling fishes that feed on the benthic infauna and epifauna.

Pelagic refers to the water columns of the ocean. The pelagic environment supports two types of marine organisms: microscopic plants and tiny animals that drift with the currents and make up the plankton, and the swimming nekton. The latter consists mainly of fishes and other vertebrates, such as turtles and marine mammals (e.g., seals, porpoises, and whales). Some invertebrate groups (e.g., squids and swimming crabs), contribute to the nekton.

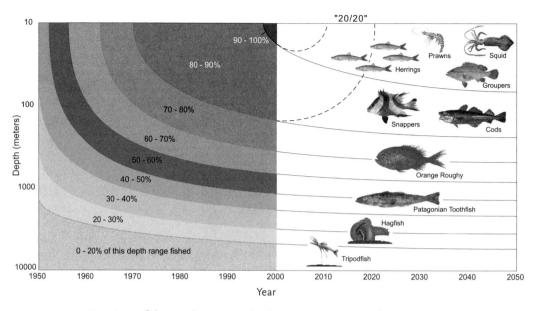

FIGURE 3. Fraction of the sea bottom and adjacent waters contributing to the world fisheries from 1950 to 2000 (*38*) and projected to 2050 by depth (logarithmic scale). Note the strong reversal of trends required for 20 percent of the waters down to 100-m depth to be protected from fishing by 2020.

1,000 m, whereas fisheries targeting oceanic tuna, billfishes, and their relatives covered the world ocean by the early 1980s (*9*).

Extrapolating the bottom fisheries trends to 2050 is straightforward (Figure 3). With satellite positioning and seafloor-imaging systems, we will deplete deep slopes, canyons, seamounts, and deep-ocean ridges of local accumulations of judiciously renamed bottom fishes—for example, orange roughy (previously "slimeheads"), Chilean seabass (usually IUU-caught Patagonian toothfish), and hagfish (caught for their "eel-skins," and here predicted to become a delicacy in trendy restaurants, freshly knotted and sautéed in their own slime). The abyssal tripodfishes are the only group that seems safe so far. Figure 3 also shows the radical trend change required to turn 20 percent of the shallowest 100 m of the world ocean into marine reserves by 2020—that is, returning to the 1970s state.

Traditional explanations of overfishing emphasize the open-access nature of the fisheries "commons." However, overcapitalized fisheries can continue to operate after they have depleted their resource base only through government subsidies (*13, 18*). Moreover, industrial fisheries depend upon cheap, seemingly superabundant fossil fuels (*19, 20*), as does agriculture. Thus, we shall here venture a prediction counter to the trends in Figure 3, based in part on the global oil production trend in Figure 4A. If fuel energy becomes increasingly scarce and expensive in the next decades as we move beyond the peak of global oil production, which is predicted to occur shortly by a number of independent geologists (*21*), then we should expect the least profitable of the energy-intensive industrial fisheries to fold (Figure 4B). This would mainly affect deep-sea bottom trawling, which drives the trends in Figure 3.

Any potential conservation benefits that might arise as a result of rising crude oil prices could be easily vitiated by increasing subsidies by short-sighted governments willing to accede to lobbying efforts from the fisheries sector. Barring escalating subsidies, one effect of high fuel prices may be to increase human consumption of small pelagics (mackerels, herrings, sardines, or

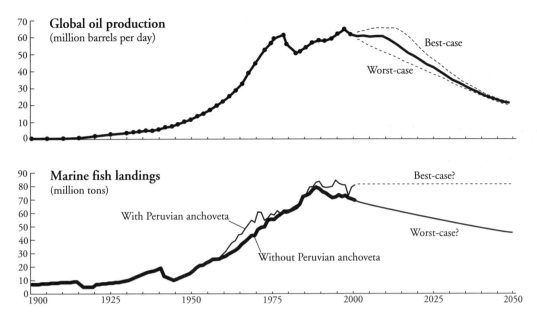

FIGURE 4. Recent historical patterns and near-future predictions of global oil production and fish catches (1900 to 2050). (A)Various authors currently predict global oil production to decline after ~2010 (21), based on M. King Hubbert's model of reservoir depletion, with worst, medium, and best cases based on different assumptions about discoveries of new oilfields. (B) Global marine fisheries landings began to decrease in the late 1980s (10). The smoothly declining trend extrapolates this to 2050 and also reflects the potential effect of future, exceedingly high fuel prices. The flat line—that is, sustaining present landings— would result from implementing proactive components of the Marketing First and Policy First scenarios.

anchovies such as the Peruvian anchoveta), which are now mostly turned into fish meal for agriculture (to raise chickens and pigs) and increasingly aquaculture (22).

However, predictions are better embedded into scenarios—sets of coherent, plausible stories designed to address complex questions about an uncertain future (23). Scenario analysis is especially important for the fisheries sector, which, although a major provider of food and jobs in many poorer countries, is small relative to the economy of richer countries and is thus "downstream" from most policy decisions.

In complement to the Millennium Ecosystem Assessment (24–27), we use the four scenarios developed by the United Nations Environment Programme (28) to investigate the future of marine fisheries. For each scenario, we also summarize results of regional simulation models

explicitly accounting for interspecies feeding interactions, within a range of ecosystem types and fisheries (29, 30).

SCENARIO 1. Markets First, where market considerations shape environmental policy. This may imply the gradual elimination of the subsidies fuelling overfishing (13). Putting markets first may also imply the suppression of IUU fishing (including flags of convenience), which distorts economic rationality in the same way that insider trading or fraudulent accounting does. Markets First, by overcoming subsidies, could also lead to the decommissioning of fuel-guzzling distant-water fleets (especially large trawlers), and perhaps lead to a resurgence of small-scale fleets deploying energy-efficient fixed gears. This scenario allows for spontaneous emergence of quasi-marine reserves (that is, areas not economically fishable, particularly

> Industrial fisheries depend upon cheap, seemingly super-abundant fossil fuels, as does agriculture.

tries, through harmful algal blooms, diseases, and invasive species.

We simulated this scenario through fleet configurations maximizing long-term gross returns to fisheries (i.e., ex-vessel value of landings plus subsidies, without accounting for fishing costs). The results are increasing fishing effort, stagnating or declining catches, and loss of ecosystem components: a large impact on biodiversity.

SCENARIO 3. Policy First, where a range of actions is undertaken by governments to balance social equity and environmental concerns. This is illustrated by the recent Pew Oceans Commission Report (32), which for the United States proposes a new Department of the Ocean and regional Ecosystem Councils, and a reform of the Fisheries Management Councils, now run by self-interested parties (33).

Similar regulatory reforms, coordinated between countries, combined with marine reserve networks, massive reduction of fishing effort—especially gears that destroy bottom habitat and generate large bycatch (34)—and abatement of coastal pollution may bring fisheries back from the brink and reduce the danger of extinction for many species.

This scenario corresponds to simulations where rent is maximized subject to biodiversity constraints. We found no general pattern for the fleet configurations favored under Policy First, because the conceivable policies involve ethical and esthetic values external to the fisheries sector (for example, shutting down profitable fisheries that kill sea turtles or marine mammals).

SCENARIO 4. Sustainability First requires a value system change to one that favors environmental sustainability. This scenario, which implies governments' ratification of and adherence to international fisheries management agreements and bottom-up governance of local resources, would involve creating networks of marine reserves and carefully monitoring and rebuilding a number of major stocks (35, 36). This is because high biomasses provide the best safeguard against overestimates of catch quotas and

offshore) and thus may reduce the impact of fishing on biodiversity. However, high-priced bluefin tuna, groupers, and other taxa (including invertebrates) would remain under pressure.

When modeled, this scenario corresponds to maximizing long-term fisheries "rent" (ex-vessel values of catch minus fishing costs). This usually leads to combinations of fleets exerting about half the present levels of effort, targeting profitable, mostly small, resilient invertebrates and keeping their predators (large fishes) depressed. Shrimp trawlers presently operate in this way, with tremendous ecological impacts on bottom habitats (29–31).

SCENARIO 2. Security First, where conflicts and inequality lead to strong socioeconomic boundaries between rich and poor. This scenario, although implying some suppression of IUU fishing, would continue "fishing down marine food webs" (6), including those in the High Arctic, and subsidizing rich countries' fleets to their logical ends, including the collapse of traditional fish stocks. This implies developing alternative fisheries targeting jellyfish and other zooplankton (particularly krill) for direct human consumption and as feed for farmed fish. This scenario, generally accentuating present ("south to north") trading patterns, would largely eliminate fish from the markets of countries still "developing" in 2050.

This scenario would also increase exports of polluting technologies to poorer countries, notably coastal aquaculture and/or fertilization of the open sea. This would have negative effects on the remaining marine fisheries in the host coun-

environmental change (*13*) (the latter is not covered here but is likely to affect future fisheries).

We simulate this scenario by identifying the fishing fleet structure that maximizes the biomass of long-lived organisms in the ecosystem. This requires strong decreases in fishing effort, typically to 20 to 30 percent of current levels, and a redistribution of remaining effort across trophic levels, from large top predators to small prey species.

These scenarios describe what might happen, not what will come to pass. Still, they can be used to consider what we want for our future. We have noted, however, that many of the fisheries we investigated—for example, in the North Atlantic (*18, 36*) or the Gulf of Thailand (*30, 31*)—presently optimize nothing of benefit to society: not rent [taxable through auctions (*37*)], and not even gross catches (and hence long-term food and employment security). It is doubtful that they will be around in 2050.

References and Notes

1. The FAO regularly issues demand-driven global projections wherein aquaculture, notably in China, is assumed to compensate for shortfalls, if any, in fisheries landings (see www.fao.org).

2. J. G. Pope, *Dana* 8, 33 (1989).

3. R. H. Parrish, in *Global Versus Local Change in Upwelling Areas*, M. H. Durand *et al.*, Eds. (Séries Colloques et Séminaires, Orstom, Paris, 1998), pp. 525–535.

4. D. Pauly, in *Ecological Integrity: Integrating Environment, Conservation and Health*, D. Pimentel, L. Westra, R. F. Ross, Eds. (Island Press, Washington, DC, 2000), pp. 227–239.

5. P. Cury, P. Cayré, *Fish Fish.* 2, 162 (2001).

6. D. Pauly, V. Christensen, *Nature* 374, 255 (1995).

7. D. Pauly *et al.*, *Science* 279, 860 (1998).

8. J. B. C. Jackson *et al.*, *Science* 293, 629 (2001).

9. R. A. Myers, B. Worm, *Nature* 423, 280 (2003).

10. R. Watson, D. Pauly, *Nature* 414, 534 (2001).

11. D. Pauly *et al.*, *Philos. Trans. R. Soc. B* 360, 5 (2005).

12. R. L. Naylor *et al.*, *Nature* 405, 1017 (2000).

13. D. Pauly *et al.*, *Nature* 418, 689 (2002).

14. T. J. Pitcher *et al.*, *Fish Fish.* 3, 317 (2002).

15. K. Kelleher, *Discards in the World's Marine Fisheries. An Update* (FAO Fisheries Technical Paper 470, 131 pp.) (2005).

16. Previous estimates for discards in the 1980s and 1990s were 27 million to 20 million tons, current estimates are 7.3 million tons; although not directly comparable, it is likely discards are lower than in earlier years (*15*). This implies that the decline of global fisheries estimated at 0.3 million tons per year (*10*) is actually two to three times higher (*17*).

17. D. Zeller, D. Pauly,. *Fish and Fisherie.* 6, 156–159 (2005).

18. D. Pauly, J. Maclean, *In a Perfect Ocean: The State of Ecosystems and Fisheries in the Atlantic Ocean* (Island Press, Washington, DC, 2003).

19. P. Tyedmers, in *Encyclopedia of Energy*, C. Cleveland, Ed. (Elsevier, San Diego, CA, 2003), vol. 2, p. 683.

20. P. Tyedmers, R. Watson, D. Pauly, *Ambio* 34, 619–622 (2005).

21. R. Heinberg, *The Party's Over: Oil, War and the Fate of Industrial Societies* (New Society, Gabriola Island, BC, Canada, 2003).

22. C. Delgado, N. Wada, M. W. Rosegrant, S. Meijer, M. Ahmed, *Fish to 2020: Supply and Demand in Changing Global Markets* (International Food Policy Research Institute and WorldFish Center, 2003).

23. G. D. Peterson, S. R. Carpenter, G. S. Cumming, *Conserv. Biol.* 17, 358 (2003).

24. The four scenarios are *Global Orchestration* (economic growth and public goods focus on a global scale), *TechnoGarden* (green technology and continued globalization), *Order from Strength* (national security and economic growth focus at regional levels), and *Adapting Mosaic* (regionalized with emphasis on local adaptation and flexible governance). The focus on alternative approaches to sustaining ecosystem services distinguishes the Millennium Ecosystem Assessment (MA; see 25) scenarios from previous global scenario exercises.

25. The Millennium Ecosystem Assessment is an international work program designed to meet the needs of decision makers and the public for scientific information concerning the consequences of ecosystem change for human well-being and options for responding to these changes. See www.millenniumassessment.org.

26. R. Watson, J. Alder, V. Christensen and D. Pauly. "Mapping fisheries patterns and their consequences," pp. 13–33., in *Place Matters: Geospatial Tools for Marine Science, Conservation, and Management in the Pacific Northwest*, D. J. Wright, A. J. Scholz Eds (Univ. of Oregon, 2005).

27. E. M. Bennett *et al.*, *Frontiers Ecol. Environ.* 1, 322 (2003).

28. United Nations Environment Program, *Global Environmental Outlook 3* (EarthScan, London, 2002).

29. This work was based on mass-balanced food web models and their time-dynamic simulation, through coupled differential equations, under the effects of

competing fishing fleets, using the Ecopath with Ecosim software, and models representing the South China Sea (with emphasis on the Gulf of Thailand and Hong Kong waters), the North Atlantic (North Sea, Faeroes), the North Pacific (Prince William Sound, Alaska, and Georgia Strait, British Columbia), and other marine ecosystems documented in www.ecopath.org and www.saup.fisheries.ubc.ca/report/report.htm.

30. V. Christensen, C. J. Walters. *Bull. Mar. Sci.* 74, 549 (2004).

31. V. Christensen. *J. Fish Biol.* 53 (Suppl. A), 128 (1998).

32. Pew Oceans Commission Report, *America's Living Oceans: Charting a Course for Sea Change* (2003).

33. T. Okey, *Mar. Policy* 27, 193 (2003).

34. L. Morgan, R. Chuenpagdee. *Shifting Gears: Addressing the Collateral Impacts of Fishing Methods in U.S. Waters* (Island Press, Washington, DC, 2003).

35. T. J. Pitcher, *Ecol. Appl.* 11, 606 (2001).

36. J. Alder, G. Lugten, *Mar. Policy* 26, 345 (2002).

37. S. Macinko, D.W. Bromley, *Who Owns America's Fisheries?* (Island Press, Washington, DC, 2002).

38. Disaggregated global landings assembled by the FAO from 1950 to 2000 were used to determine when each 30 min by 30 min spatial cell was first "fished" [i.e., when landings of fish (other than oceanic tuna and billfishes) from that cell first reached 10 percent of the maximum landings ever reported from that cell]. The percentage of cells fished at each depth was then calculated.

39. D.P., J.A., V.C., and R.W. are members of the Sea Around Us Project, initiated and funded by the Pew Charitable Trusts, Philadelphia. We thank A. Kitchingman and W. Swartz for help with the figures.

Web Resources

www.sciencemag.org/cgi/content/full/302/5649/1359/DC1

Physical Resources

Preserving the Conditions for Life

DONALD KENNEDY

Human populations and their rates of growth, biodiversity, and the health of marine ecosystems present key scientific challenges for understanding how to sustain life on earth. But the physical environment sets the conditions for life. The physical environment includes:

- the soils that determine agricultural productivity, especially in the developing world;
- fresh water, an essential for direct human use and for agriculture and the support of natural ecosystems;
- energy, used by our growing human population to support economic development, often at environmental risk;
- air quality, with its effects on human and ecosystem health; and, finally,
- the contemporary problem of climate change and the factors that drive it.

Vulnerability and Diversity in Tropical Soils

Soils in general, and tropical soils in particular, are some of the most vulnerable resources on the planet, because of human overharvesting and misuse and also because of environmental change. Tropical soils, like all soils, are critical to the world's food security—the guarantee that all people have physical and economic access to enough safe and nutritious food to meet their dietary needs. Many readers will have heard this problem analyzed in terms of world supplies of wheat, rice, corn, and a few other staple grain crops that are traded in global markets as well as grown directly to supply farmers in temperate zones. But the tropics are different. They contain the world's poorest people, many of whom are rural, low-income farmers who supply their families (and perhaps their small communities) with locally grown food. In general, they lack the incomes necessary to supplement what they grow themselves with food purchased from outside sources.

For these people, the statistics showing that the yields of the "big five" major grains have more than kept pace with global population growth have no meaning. They have more pressing problems to worry about. As Stocking points out, the soils in many tropical locations lose their value rapidly in agricultural use—even under good soil management practices. Tropical soils are highly variable: some are especially sensitive to erosion and nutrient loss, whereas others—such as the upland soils of East Africa—are not. In addition to their sensitivity, differences in their resilience—their capacity to recover and improve yields—are important. Even some sensitive soils can be resilient provided that good management is applied during recovery.

Preparing for Change

The current status of tropical agriculture varies from one agroclimatic zone to another, but the aggregate picture suggests a cumulative loss in soil quality. What can be done to improve the picture? First, there are clear successes in soil conservation, and the lessons they provide suggest that there are effective ways of improving local management, such as new techniques in fallow cropping or intercropping. Where refugees or immigrants enter an area, local networks can improve the success rates of the new farmers. Indeed, community engagement to share "best practices" among rural smallholder farmers may represent the best hope for improvement.

Second, it will be necessary for tropical agriculture to be prepared for change. Unfortunately, the status quo is not going to be with us for very long. Continuing increases in average global temperature have already changed the character of some agroclimatic zones. As global warming progresses, two outcomes for tropical agriculture seem likely if not certain. First, the tropics will be affected by the temperature change itself. Even though the largest temperature changes will occur at higher latitudes, tropical croplands and tropical farmers have little capacity for adaptation: soils are vulnerable and economic limitations are stringent. Many tropical plants and animals tend to be closely adapted to an optimal temperature, suggesting that temperature change may lead to general ecological disruption. Finally, climate change models project an increase in the frequency of "extreme weather events," and tropical farmers have come to rely on fairly regular patterns of temperature and precipitation cycles. Flooding and landslides can devastate productive tropical agriculture, and have already done so in Central America and southern Africa.

Orphan Crops

A major problem for the future of tropical soils and food security is the extent to which global research efforts have been focused on the major grain crops while the food plants that are critical for regional survival have been relatively ignored. In "Tropical Soils and Food Security: The Next 50 Years," Stocking mentions the extraordinary diversity of yams (54 varieties!) that are cultivated in Ghana. There, local knowledge is apparently effective in matching these varieties to particular soil conditions. But yams, millet, taro, sweet potatoes, and plantain are examples of "orphan crops" that have received little attention from plant breeders and geneticists. For some of these crops, it is possible to identify only a few experts worldwide who specialize in their genetics and agronomy. There could be important gains for the world's poor if developed-country governments and other institutions would enlarge their scientific investments in these "orphan crops."

Water: Access, Allocation, and Quality

Moving from tropical agriculture to water is not as sharp a transition as one might think. Not only do growing plants need water, but it is among the nonliving resources critical to all life—including the human variety. Water may be readily available in the humid tropics, but it is notably scarce in many drier low-latitude habitats. Even in India, the epicenter of the Green Revolution, farmers must pump groundwater from several kilometers down to water their crops. But in the arid temperature zones, agricultural water makes huge demands on limited supplies. In California, for example, huge projects divert water from the north Sierra to the heavily populated southern part of the state. The political claims of California's Central Valley agriculture over the years have resulted in water delivery at a heavily subsidized price, with the result that municipal and industrial users are essentially financing the irrigation of farmlands and leaving nonagricultural users in shorter supply.

Water and Big Projects

As Gleick shows in the chapter "Global Freshwater Resources: Soft-Path Solutions for the 21st Century," the classic solution to the problem of water availability has entailed reliance on "hard" infrastructure: for example, the construction of dams, aqueducts, and other major projects to transport water from one drainage basin to another, and the canalization of rivers. Although these projects have improved water availability in a number of cases, they have also produced an array of adverse human and environmental consequences. The displacement of Chinese residents in connection with the Three Gorges Dam and reservoir project, the declining availability of water for Mexico in the Colorado River delta in the Sea of Cortez, and the California transfer system are particularly dramatic cases. Yet as Gleick argues, although these projects have made more water available in terms of total withdrawals, 1 billion people still lack access to safe drinking water, and more than twice that many are not served by sanitation facilities. The consequences for human health are enormous.

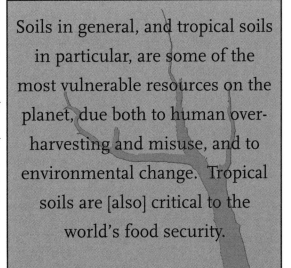

Soils in general, and tropical soils in particular, are some of the most vulnerable resources on the planet, due both to human over-harvesting and misuse, and to environmental change. Tropical soils are [also] critical to the world's food security.

A Softer Approach

This section makes a strong case for a "soft path" in which economics plays a significant role. Its objective is not simply to produce "more water"; it is to produce more satisfaction on the part of consumers who use water to fulfill basic human needs. A householder in the United States, if asked to rank his water needs, might produce something like this: first, enough to drink to be adequately hydrated for health; second, enough to take a shower; third, enough to take care of sanitation; fourth, enough to water the lawn and garden. His willingness to pay would be highest for that first unit volume of drinking water, and least for the lawn. But production economies (improved flush-toilet technology, drip irrigation for the lawns) could lower the costs for those lower-ranked uses, allowing more for the more vital uses.

Consumer behavior drives technology development. Gleick points out that the amount of water required to flush all the U.S. toilets has dropped by almost 75 percent over the last two decades. Gradually, the incentive structure in the U.S. economy has begun to take hold in the water sector. A given volume of water today generates nearly twice as much gross national product as it did at the turn of the 20th century. That means that the economic efficiency of water production and use is headed in the right direction, along the "soft path."

Economies and Incentives

Returning to California water, however, there is a problem the author of the chapter knows well. He points out that drip irrigation, soil conservation, crop selection, and a variety of other techniques can increase the economic value produced by a given volume of water. Good farmers are adopting these techniques, thus making more water available to others. But why does one often see furrow irrigation or wasteful overhead spraying in Central Valley farms? Farmers in California pay far less than industrial or municipal users; they are getting water at a subsidized bargain price. That lessens their incentive to save money by using less of it, so they may decide not to make capital investments in a drip system. As economic critics of this subsidy often say, "to make it work, you've got to get the prices right." In many places the price of water, with its complex political history, just isn't right.

Later in this collection, you will read about the "Tragedy of the Commons." Garret Hardin's famous article in *Science*, now 38 years old, pointed out that if there is a common pool resource, individual users have an incentive to take an extra amount because their excess consumption comes largely at the expense of others who share the resource. Water in this case is not a free good, but if it is available at an unusually low price, consumers will treat it like one. Think about subsidized water as a "commons," and ponder how the problem might be corrected.

Energy: Another "Soft Path" Candidate?

From water to energy is only a short trip, and the trail along which we will travel to connect the two is what we have been calling the "soft path." Like soil and water, energy is among the resources most critical to human well-being. Like soil and water, energy is also scarce, not to mention fraught with economic and political perils. The authors of the chapter "Energy Resources and Global Development," give a thoughtful and extensive summary of energy use—by country (developed/developing), by economic sector (transportation, residential, commercial), and by type (coal, oil, gas, "renewables"). Their conclusion, surely now a consensus in the scientific community, is that our dependence on fossil fuels will continue for some time. That poses problems for access, since the number of producing nations is limited and geographically scattered. In turn, that acts as an incentive for nations that lack adequate reserves (such as Japan) to pursue geopolitical arrangements to guarantee supply and practice energy conservation at the same time.

Energy, Efficiency, and Development

That the developed nations continue their dependence on fossil fuels does not mean that they have not made some improvements. "Energy efficiency"—that is, the value of national product generated by one unit of energy—has increased in most developed countries, although some, such as Japan, have reached much higher levels than others. Developing nations, where the major use is residential, have shown fewer gains and continue to be more dependent on biomass burning. But these countries have an appetite for economic development, and as their transportation needs grow they will have more need for fossil fuels. An important question is this: What will the developing-world energy trajectory look like? As far as energy efficiency is concerned, it is likely to follow the pattern already set by the industrial nations. But a different issue raised by the authors has to do with the environmental impact of development. Are these nations doomed to repeat the painful and environmentally dirty transition of Victorian economies into the postindustrial world? Or will new technologies allow them to leapfrog over the more difficult parts of the transition and thus move along a softer path?

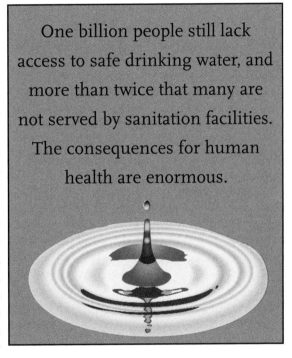

One billion people still lack access to safe drinking water, and more than twice that many are not served by sanitation facilities. The consequences for human health are enormous.

Of course one of the reasons for worrying about this connects with a later chapter in this book, dealing with climate change. The primary source of increases in atmospheric greenhouse gases, and thus of increasing global temperature, is the combustion of fossil fuels. Most of the carbon dioxide, the major greenhouse gas, has come from the developed countries—particularly the United States. It is thus in the interest of the United States to slow its own emissions, and thus by example encourage the developing countries to follow suit. But to maintain anything like our current profile of national economic activity, substitute sources of energy must eventually be found to replace oil, coal, and natural gas.

Changing Energy Sources

The authors of "Energy Resources and Global Development" explore a number of possibilities, and the result does not give much reason for encouragement. For most renewables (wind, solar, geothermal) the cost factors are not promising, and there are some environmental problems with each. Nuclear energy, because of mounting public concern over global warming, is attracting more attention because it is carbon-free. But the problem of diversion of fissile materials and the unresolved challenge of how to store high-level nuclear wastes continue to stand in the way of significant new nuclear development, and in most industrial nations save France political objections have been strong enough to discourage adoption. Given our present commitments to fossil fuels for electric power production, transportation, and industry, it seems unlikely that there will be a major transition to a new energy economy any time soon. But of course external factors could change that. If cost increases (through taxes or market decisions by oil-producing countries) were high enough, an economic advantage could appear for some alternative energy sources. Alternatively, increased recognition of the hazards of climate change could change both the political and economic landscape for energy sources that are carbon-free.

Air Quality as a Global Problem

Global air quality and pollution, whose importance became so clear only in the past few decades as satellite observations began to map the worldwide distribution of air pollutants, is considered by Akimoto in the chapter "Global Air Quality and Pollution." The revelations resulting from these new measurements illustrate the widespread occurrence of major pollutants, and—remarkably—the degree to which they are moved from one continent to another by upper-level winds. Atmospheric chemists in the northwestern United States now speak regularly of the transport of pollutants from Asia as the "trans-Pacific express." But, as Akimoto shows, the intercontinental movements involve many continents and many directions: Europe to Asia and North America to Europe, for example.

These cases of distant transport are important for many reasons. Some of the transported pollutants may have human health significance: carbon monoxide and ozone, for example, present real risks at the level of the troposphere—the portion of Earth's atmosphere that is down where people live. Aerosols—formed by oxides of sulfur and nitrogen, carbon-containing dusts, and airborne sea salt—have significant influence on climate, producing a cooling effect that counteracts the "radiative forcing" exerted by the greenhouse gases. Major volcanic eruptions can cause huge emissions of these aerosols: after the eruption of Mt. Pinatubo in 1982 average global temperature was significantly lowered for 2 years.

Some of these aerosols, such as sulfur dioxide (SO_2), have created international tensions because of their transport across boundaries. In the 1980s it became apparent that SO_2 produced by factories and power plants in the Midwestern United States were producing adverse effects on forests and waters in the Northeast and into Canada. Thrown into higher-altitude winds by tall stacks, this compound—readily converted to sulfuric acid in water—was falling onto distant lakes and woodlands as "acid rain." Protests were lodged by the Canadian government, and after a long and politically tense series of negotiations, a U.S. government survey was done in the late 1980s. Neither it nor attempted Congressional action did anything to solve the problem, but in the 1990 Clean Air Act Amendments, the United States adopted a program of tradable emission permits that had begun to limit the SO_2 release from sources in that country.

As developing nations increase industrial activity and add new sources of pollution to biomass burning, air-quality problems are likely to worsen. The figure showing nitrogen oxide emissions from three continents—Europe, North America, and Asia—demonstrates this trend dramatically. In 1980, Asian emissions had less than half the emissions of the other two, but in 20 years it has caught and passed both the others. As Asia grows and develops, it will increasingly be dominated by what Akimoto calls "megacities"— huge concentrations of people with populations of 10 million or more. The number of megacities is now well over 20, and it may be necessary to invent a new size category— perhaps "gigacity"—for those with populations over 25 million. These places not only represent health threats to their own citizens; in addition, their emissions may contribute significantly to regional or even more distant climates.

The Changing Climate

Climate change, one of the most pressing and troublesome global problems facing humanity, is the topic of the final chapter in this section.

Garret Hardin's famous article in *Science,* now 37 years old, pointed out that if there is a common pool resource, individual users have an incentive to take an extra amount because their excess consumption comes largely at the expense of others who share the resource. Water in this case is not a free good, but if it is available at an unusually low price consumers will treat it like one.

It is written by Thomas Karl and Kevin Trenberth, two climate scientists who are associated with national centers—the National Oceanic and Atmospheric Administration (NOAA) and the National Center for Atmospheric Research (NCAR)—that compile and study climate data and perform atmospheric research. Their chapter represents the studied consensus of the scientific community, and its conclusions reflect, in summary form, most of those reached by the Intergovernmental Panel on Climate Change. That organization, which involves over 200 scientists from around the world, was put in place by the World Meteorological Organization and the United Nations. Karl and Trenberth conclude that human activities have raised the average global temperature by about 0.7 degrees centigrade. This is due mostly to the emission of greenhouse gases such as carbon dioxide, caused mostly by fossil fuel burning over the past 100 years, especially during the past 40 years, whose concentration in the atmosphere has risen from a pre-industrial value of 280 ppm to around 370 ppm. Other anthropogenic emissions, including those of methane, nitrous oxide, and soot, have also contributed to the warming of the near-surface atmosphere of the planet.

Climate Predictions

Climate models, as the authors show, predict that further increases in greenhouse gas emissions will continue to raise the average temperature of the globe. A warmer Earth means more evaporation, and this intensification of the hydrological cycle raises the probability of extreme weather events. Will we continue to experience a slow ramp of increasing global temperature? That will depend on several future events or actions that cannot be predicted with certainty. If there were abrupt decreases in carbon dioxide emissions, that would slow the process, but of course it is uncertain whether such major changes in the behavior of individuals or societies will actually take place. The authors describe a number of "feedbacks" that could modify the rates. The process through which average global temperature is being increased by the addition of greenhouse gases could be accelerated—as it is now in many places—by the melting of ice to expose bare ground, which absorbs heat instead of reflecting it. It could be slowed temporarily by any massive injections of aerosols, as from significant volcano activity. Changes in cloud cover could either increase or decrease rates of heating. Finally, some scientists believe that major releases of fresh water in the North Atlantic region could interrupt the process by which heat is brought from the equator northward in the Gulf Stream. This northward current is half of a circuit that must be maintained by a deep current of cold, dense water. To produce it, salinity increases and cooling as the current bends to the east must make the water mass dense enough to sink. According to this hypothesis, freshwater injections from melting ice and failure of evapora-

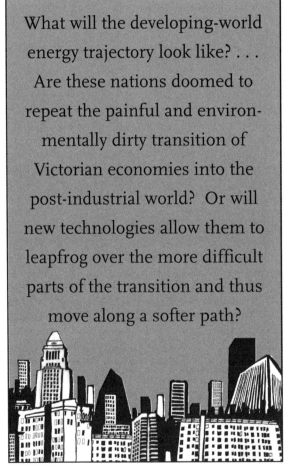

What will the developing-world energy trajectory look like? . . . Are these nations doomed to repeat the painful and environmentally dirty transition of Victorian economies into the post-industrial world? Or will new technologies allow them to leapfrog over the more difficult parts of the transition and thus move along a softer path?

tion would prevent sinking, and as a result the system could "stall" and make Europe much colder.

These future uncertainties have encouraged some who deny that climate change represents a significant problem. But the influence of the heating we have already observed should be enough to raise concerns. Glaciers everywhere are thinning and melting. Flowers are blooming weeks earlier in many regions, and the breeding cycles of birds have been altered so that reproduction begins much earlier. Unusually hot summers, regional droughts, and flooding and landslides due to exceptionally heavy rain have been much more frequent during the last decade than previously. These effects have been subjected to more intense scrutiny since the article by Karl and Trenberth was published, and their

> Climate models . . . predict that further increases in greenhouse gas emissions will continue to raise the average temperature of the globe. A warmer Earth means more evaporation, and this intensification of the hydrological cycle raises the probability of extreme weather events.

likelihood has been repeatedly confirmed. Other recent results have emphasized that much of the excess heating we have already experienced has been transferred into the world's oceans, which will give it up slowly. That represents a commitment to climate change that will continue to affect the Earth even if the emission of greenhouse gases were to be stabilized. Finally, the authors point out, we will not be able to address the challenges with which climate change will present us without greater international cooperation and action.

Climate History

The history of climate change has been worked out through difficult and technologically demanding experiments on cores of ice obtained from glaciers in Greenland and Antarctica, and from the analysis of marine sediments. What these tell us about the geologically "recent" history of Earth's climate is an important part of the human narrative. For most of the past 110,000 years, we know that the climate fluctuated between "warm" states resembling the present regime and prolonged "cold" states marked by glacial advances and temperatures 8°C or more below our present average. These changes, tracked in the ice cores through oxygen isotope ratios that provide a reliable proxy for temperature, were probably induced by a variety of influences, including changes in Earth's orbit around the sun. Many were surprisingly rapid, sometimes changing the average temperatures in a region by 10 degrees or more in as little as 20 years.

It is interesting to imagine how our ancestors dealt with a climate like that. At the time of the Last Glacial Maximum, 20,000 years ago, so much water was tied up in continental glaciers that present Chicago was under 3 kilometers of ice, dropping sea level by so much that one could walk from London to Paris had those cities existed then. Yet our Paleolithic ancestors were creating beautiful paintings in the caves of Spain and southern France, making religious objects, and subsisting effectively as hunters and gatherers. After that the climate warmed somewhat, but about 13,000 years ago there was a last blast of glaciation—the event referred to in the chapter as the *Younger Dryas*.

Then an astonishing thing happened. The climate record leveled out, and since about 11,500 years before the present time we humans have lived in a stable, benign period unlike any known in the last 120,000. The Holocene, as it is called by geologists, has had minor shifts in climate—the Little Ice Age and the Medieval Warm Period—but these cannot be compared with the huge, dynamic changes that had gone before. It is not surprising that human agriculture and permanent human settlement originated near the beginning of the Holocene. Thus we can attribute much of what is modern and familiar about human societies and their creations to our good fortune in having been given this ten-millennium break in the climate.

The stability of our climate is the predicate for the state of all the other resources, living and nonliving, on which we depend. So, in a sense, the job of a variety of scientists—atmospheric chemists, paleoclimatologists, oceanographers, and others— is to learn what influences are at work that could take us out of the present gentle regime. Because the nature of our physical Earth is so vital to the conduct of human activity, climate change and what to do about it has become one of the great international political struggles of our time. It is now clear that the climate regime has been changed as a result of human activity; in fact, a few geologists have begun to label the last 100 years of the Holocene the "Anthropocene" to signify that for the first time humans are in control of the physical state of the planet. The great question is: When will the Anthropocene end, and how?

Tropical Soils
and Food Security
The Next 50 Years

MICHAEL A. STOCKING

An appreciation of the dynamism of the links between soil resources and society provides a platform for examining food security over the next 50 years. Interventions to reverse declining trends in food security must recognize the variable resilience and sensitivity of major tropical soil types. In most agroecosystems, declining crop yield is exponentially related to loss of soil quality. For the majority smallholder (subsistence) farmers, investments to reverse degradation are primarily driven by private benefit, socially or financially. "Tragedy of the commons" scenarios can be averted by pragmatic local solutions that help farmers to help themselves.

The UN Food and Agriculture Organization (FAO) defines *food security* as "when all people, at all times, have physical and economic access

to sufficient, safe and nutritious food to meet their dietary needs and food preferences for an active and healthy life" (1). Currently, more than a billion people have no food security. About 60 percent of rural communities in the tropics and subtropics are persistently affected by decline in household food production, with sub-Saharan Africa and parts of Latin America, the Caribbean, and Central Asia suffering worst (2–4). Technology, such as irrigation and improved crop varieties, has changed the situation for some people, but insecurity still prevails for the poorest and most vulnerable. The 1996 World Food Summit in Rome aimed to cut by half the number of undernourished people in the low-income food deficit countries (5). The target was reaffirmed in 2000 for achievement by 2015 in the Millennium Development Goals (6).

Reviews of techniques to improve soils through conservation and better management are common [e.g., in Central America (7)]. Food security requires policy adjustments in science,

This article first appeared in *Science* (21 November 2003: Vol. 302, no. 5649). It has been revised and updated for this edition.

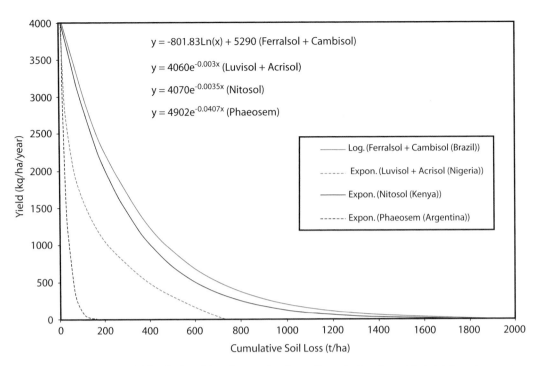

FIGURE 5. Erosion-yield relationships for a selection of tropical soils, with initial maize grain yield on virgin land set to 4000 kg ha^{-1} [Source: (60)]. The initial yield level is artificially set at 4 tonnes on the basis of the amount of maize required by one typical household for 1 year in the subhumid tropics. Yields under subsistence agriculture are often well below this level. Exactly the same relationships hold true for lower starting levels, except that the time to reach critical food insecurity will be shortened, hence the serious challenge to agriculture and soil management.

organization of science, and engagement with farmers rather than revolutionary change (see Figure 5). The headline challenges for attaining global food security are political [e.g., conflict over land (8)], climatological [e.g., drought and global warming (9)], or epidemiological [e.g., impact of AIDS/HIV on farm labor (10)]. All too rarely is the underlying and changing quality of the natural resource base advanced as a key determinant of the increasing vulnerability of poor people to food insecurity. So what is the evidence that changing soil quality reduces food production? How may we better intervene to provide food security in the next 50 years?

Soil Quality

Soil quality is "the capacity of a soil to function within land use and ecosystem boundaries, to sustain biological productivity, maintain environmental quality and promote plant, animal and human health" (11, 12). As a concept, it differs from traditional technical approaches that focus solely on productive functions, including soil, water, and chemistry.

Instead, soil quality is a holistic concept that recognizes soil as part of a dynamic and diverse production system with biological, chemical, and physical attributes that relate to the demands of human society (13, 14). Society, in turn, actively adapts soil to its needs, mining it of its nutrients on demand and replenishing these nutrients in times of plenty.

Diverse interactions among soil, its productive output, and society are involved. These interactions include the outcomes of combinations of plant nutrients (15), the complex processes of change in capturing carbon (16), and the potential positive effects of soil amendments such as green manure on crop productivity (*17, 18*).

Assessing soil quality is a major challenge because quality varies spatially and temporally and is affected by management and the use of the soil resource. Integrated biological soil management is now recognized as essential, both for productive efficiency and for maintaining biodiversity (*19*). For example, traditional forms of conserving biodiversity in Ghana mean protecting at least 54 varieties of yam (*Dioscorea* spp.) while also managing the soil in sensitive ways that maintain soil-plant relationships. A product of this management is that the natural forest vegetation is kept to protect the climbing yams (*20*). Soil quality is a concept well understood by local farmers in this forest savanna zone, and new assessment methods stress the importance of adopting a strong farmer perspective (*21*).

Situations such as these are common throughout smallholder farming in the tropics [e.g., (*22*)]. Soil management is an intrinsic part of the overall management of biodiversity—now called *agrodiversity* (*23*)—and is therefore strategically a component of world development issues such as food security for growing populations and the provision of environmental services (*24*).

The Evidence of Impact

Depletion of soil nutrients—just one indicator of declining soil quality—is a major cause of low per capita food production, especially in Africa. Over the past 30 years, annual nutrient depletion rates for sub-Saharan Africa have amounted to a fertilizer equivalent value of US$4 billion (*25*). However, ascribing a decline in food production unambiguously to the effects of changing soil quality is difficult because of the complex interactions involved. Yields decline for many reasons, such as excessive off-take of nutrients in crops without replenishment, pests and diseases, weed infestations, and increasing prevalence of climate change–induced drought (*26*). There is an emerging understanding of the importance of microbial communities for soil health through the use of DNA and RNA methods to determine physical and chemical changes in soil (*27*). Many soil factors are involved, including soil depth

The UN Food and Agriculture Organization defines food security as "when all people, at all times, have physical and economic access to sufficient, safe and nutritious food to meet their dietary needs and food preferences for an active and healthy life." Currently, more than a billion people have no food security.

and rooting, available water capacity, soil organic carbon, soil biodiversity, salinity and sodicity, aluminum toxicity, and general acidification.

The Soil Quality Institute has suggested several indicators that relate to yields (*28*), from which indexing approaches to measure soil quality have been devised (*29*). One of the main factors that integrate the effects of others is reduction in soil organic carbon (*30*). But these indicators do not offer a comprehensive measure of the spatial and temporal variability of soils, nor

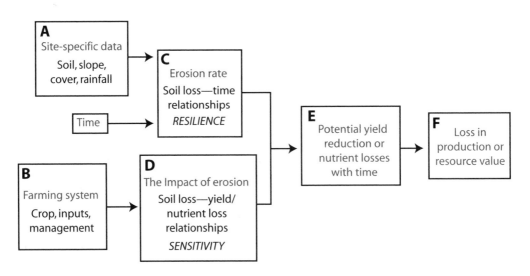

FIGURE 6.   Conceptual framework for modeling erosion-yield-time relationships.

of the dynamic relationships of soils and the people who work them (31). This lack has led to calls for interdisciplinary studies to understand how soil properties and processes interact within ecosystems (32); how economic utility is affected (33); and how society, culture, and local knowledge are influenced (34). Two topics that have attracted considerable attention for tropical agroecosystems are (i) the effect of declining soil quality on production, and (ii) the rationality and private benefit of farmers' investments in soil conservation.

The cumulative loss of productivity from soil degradation of virgin land in all agroclimatic zones is estimated to be 5 percent (35). But this estimate hides large differences between zones and the vulnerabilities of some soils in the tropics (36). Since 1984, a series of experiments on major tropical soil types has attempted to determine the relationships between crop yield and cumulative soil loss (37). Results [e.g., (38, 39)] indicate that yield decline follows a curvilinear, negative exponential that holds true for most tropical and many temperate soils (Figure 6). Erosion selectively removes the finer and more fertile fraction, with enrichment ratios of eroded sediments being highest on virgin soil (40). As the soil progressively loses its quality, subse-

quent erosion has less proportional impact. Although productivity diminishes rapidly in the early stages of erosion, different soils show different degrees of impact after varying amounts of time and prior erosion (41).

Understanding these patterns of yield decline is crucial to determining the implications of changing soil quality on future yields and food security. But can tropical smallholder farmers access the resources to maintain soil quality in the face of these trends?

Resilience and Sensitivity

It is important to distinguish the intrinsic susceptibility of the soil to erosion (its *resilience*) from the variable impact of that erosion on yields (its *sensitivity*). These concepts for soil management combine through changing soil quality to affect actual production (Figure 7). Resilience includes the "strength" of the soil or its resistance to shocks such as severe rainstorm events. Sensitivity denotes its fragility or its susceptibility to decline in production per unit amount of degradation (42). Resilience is manifested through specific degradation or erosion rates on different soils subject to the same erosive conditions, whereas sensitivity is a measure of how far the

SENSITIVITY

	High	Low
High	Resists erosion well but, if allowed to degrade through poor management, yields and soil quality rapidly decline. Nevertheless, capability can be restored by appropriate land management [Phaeozems]	Only suffers degradation under very poor management and persistent mismanagement. Biological conservation methods adequately maintain production and protect soil [Nitosols; Cambisols]
Low	Easy to degrade, with disastrous effects on production—loss of nutrients and plant-available water capacity. Difficult to restore, and should be kept under natural vegetation or forestry [Acrisols]	Suffers considerable erosion and degradation without apparent influence on production or overall soil quality. Physical structures indicated to control off-site impacts [Ferrasols]

RESILIENCE (left axis)

FIGURE 7.   A resilience-sensitivity matrix for tropical soils (using FAO/UNESCO Soil Map of the World names), with indicated approaches to conservation. Under this schema, the top right cell represents the best soils and the bottom left the worst. Different management strategies for each of the four permutations are indicated, which in turn affect the extent to which smallholder farmers have the resources and understanding to cope with specific scenarios of soil degradation (61).

change induced in soil quality affects the soil's productive capability.

A simple matrix (Figure 7) illustrates the possible permutations of resilience and sensitivity for a few tropical soils. The management strategies adopted for different permutations of sensitivity and resilience can thus be related to the capacity of local smallholder farmers to take remedial action. For example, ferralsols (35 percent of the tropics and subtropics) and acrisols (28 percent), typical of humid rainforests and shifting cultivation, have low to very low resilience and moderate sensitivity. Once vegetation is removed, they degrade quickly and irreversibly through intense acidification, increasing free aluminum and phosphorus fixation. Combina-

tions of structures such as terraces and biological measures such as intercropping are the most effective response. Mechanical structures require human and financial resources, especially labor and money, and are out of the reckoning of most subsistence farmers.

Nitosols (3 percent), by contrast, have moderate resilience and low sensitivity. They are typical of highland clay-rich areas, such as Ethiopia and Kenya, and are one of the safest and most fertile soils of the tropics and subtropics, with only minor problems of increased erodibility if organic matter declines.

Biological conservation methods are effective ways to address both erosion rate and fertility decline. They include a wide array of techniques

SCIENCE IN THE NEWS

Science, Vol. 304, no. 5677, 1616–1618 , 11 June 2004

WOUNDING EARTH'S FRAGILE SKIN
Jocelyn Kaiser

The rebel insurgency that toppled Haiti's government in 2004 was, on the surface, simply the latest turn in a bloody cycle of repression and reaction. But deeper down, the uprising exposed a root cause of the country's misery: some of the most denuded land on the planet. Only 3 percent of the once lushly forested terrain still has tree cover, and up to one-third, some 900,000 hectares, has lost so much topsoil that it is no longer arable, or barely so. Today the landscape is crisscrossed by gully-scarred, deforested hills and "rocks where there used to be dirt," says Andy White, an economist at Forest Trends, a nonprofit organization in Washington, D.C. Erosion has brought the Caribbean country's agriculture to its knees, he says, and that in turn "has driven rural poverty," impelling desperate Haitians toward city slums where unrest continues to simmer.

Sore spot. Erosion from timber cutting, overgrazing, and other human activities can leave land irreversibly damaged.
CREDIT: GRAEME EYRE/ISTOCKPHOTO.COM

The tremendous hemorrhaging of fertile soil makes Haiti one of the global hot spots for degradation, as illustrated on the preceding pages. For decades soil scientists have deplored the loss of land to erosion, nutrient depletion, salinization, and other insults. But policymakers have been slow to respond. One reason is that dire warnings in the 1980s that soil erosion would doom the world to chronic food shortages have failed to pan out.

But in the past few years, new studies have yielded better data linking soil degradation to a slump in the growth of global crop yields, uniting dueling camps. After years of "pecking away at this problem," soil scientists, geographers, and economists "are coming toward the middle," says Keith Wiebe, a resource economist at the U.S. Department of Agriculture (USDA). The bottom line: "As a global problem, soil loss is not likely to be a major constraint to food security," says Stanley Wood, a senior scientist at the International Food Policy Research Institute in Washington, D.C. But as Haitians know all too well, degradation can be a severe and destabilizing threat in places where farmers are too poor to curb or overcome the damage.

Moreover, global warming is expected to exacerbate regional crises. As the ground heats up, organic matter decomposes more readily, reducing soil fertility and, asserts Duke University ecologist William Schlesinger, releasing carbon dioxide into the air to fuel warming. Deserts are also expected to expand and erosion worsen if violent storms occur more frequently. But there's a flip side, says soil scientist Rattan Lal of Ohio State University, Columbus: "We can do something about it" by managing soils to stem erosion and retain more carbon.

involving the management of biomass: crop residues, green manures, and alley cropping, for example. The principal limitation here is the availability of organic resources and the human resources to manage them for effective soil protection. However, even at moderate levels of management such as those available in most smallholder farming households, nitosols can effectively continue to withstand degradation and produce indefinitely, at least for the next 50 years (43).

Appropriate Responses to Changing Soil Quality

Doomsday scenarios of increasing population and declining soil resource quality fail to capture the diversity of soils, while presenting the worst-case outliers as the typical situation. There are indeed some bad cases where degradation is rife and people are starving (44); these will continue to hold international attention for at least the next 50 years. International attention will also continue to be captured by global assessment initiatives such as the Millennium Ecosystem Assessment (45) that present pessimistic scenarios of the future, or initiatives to sequester more carbon into soils for gains in both global environmental benefits and food security, the "win-win" beloved of the World Bank (46). However, smallholder farmers in the tropics are the ultimate solution. They turn scenarios into futures. Their skills and social networks give us cause for optimism for future soil quality and food security. Many are already managing their soils sustainably and productively.

Although these smallholder farmers tend to be limited in labor resources (i.e., have low "human capital"), they compensate in forms of collective action and networking (i.e., "social capital") (47, 48). This means that they adapt technologies to their local needs (using indigenous knowledge and innovation) and avoid labor-demanding and expensive practices. Interventions that use community-based approaches that empower farmers to manage

their own situations therefore hold the greatest promise for maintaining soil quality and ensuring food security.

One of the most interesting recent developments in soil quality research, as reflected in the agendas of the major development agencies, is the recognition that farming practices do not merely extract soil nutrients, but they evolve in response to changing conditions over many years through informal experimentation and experience (49). Farmers may often make better decisions than the "experts," not because of any greater analytical skills but because of the experience gained in integrating a vast array of local factors responsible for controlling production.

Farmers are unlikely to undertake practices that undermine the future and put household livelihood and food security at risk unless immediate survival is in question. They will invest in soil conservation if the private benefits—financial, social, and cultural—are greater than the costs.

The greatest damage to soils occurs where conditions are volatile with, for example, migrants and refugees. Here, local knowledge is poor and mining the soil of its nutrients is essential for sur-

> Smallholder farmers in the tropics are the ultimate solution. They turn scenarios into futures. Their skills and social networks give us cause for optimism for future soil quality and food security. Many are already managing their soils sustainably and productively.

vival, at least in the short term. The greatest threat to soil quality and food security occurs if the security of tenure for smallholders is made even more difficult by changing world conditions.

The future rests in managing changes in soil quality through working with local communities and rethinking how soils change local society (50). In subhumid Benin, for example, *Mucuna pruriens* (velvetbean) is used as a green manure mulch and has become an accepted method for countering soil fertility decline.

From 15 original farmers involved in experimentation and adaptation in 1987, 100,000 were reported to have embraced this practice by 1996 (51). This is evidence of adaptability, flexibility, and responsiveness to techniques that bring private benefits to smallholders. In semi-arid Kenya, farmers choose "trashlines" (bands of uprooted weeds and crop residues) to intercept sediment and runoff, a technique never promoted by the advisory services. Yet when the marginal rates of return and net present values over 10 years are calculated, trashlines are almost always the only technique of soil quality maintenance that consistently benefits the farmers' livelihoods (52).

Such findings point to some important, if uncomfortable, conclusions. Many farmers in the tropics are willing to invest in the future, protecting important public goods such as soils, and they are often the best arbiters of choice when it comes to technologies. Science does not always get it right and does not necessarily provide workable or acceptable solutions (53). Soil resources are not a static, homogenous medium; they are a dynamic element, responding to the demands placed on them by human beings and governing expectations of food security. If simple provisions—such as adequately resourced extension services and access to technologies—are made available, food production by smallholders can be transformed. The "tragedy of the commons" lies more in our simplistic, linear, disciplinary thinking than in reality. The challenge is to capture diversity by developing appropriate cross-disciplinary analytical methods and

measures. A reinvigorated science and technology for agricultural development is necessary, but it is not sufficient. We must change mindsets that solutions are already available; the change is in how and to whom they are available. We must get closer to allaying the tragedy by providing realistic and accessible interventions for those who need it most—the poor, hungry, and disadvantaged, living on soils that are sensitive and lack resilience.

References and Notes

1. United Nations Food and Agriculture Organization, Special Programme for Food Security (www.fao.org/spfs).

2. "Partnerships to fight poverty," *Annual Report 2001* (United Nations Development Programme, New York, 2001).

3. "Reaching Sustainable Food Security for All by 2020: Getting the Priorities and Responsibilities Right," slideshow presentation (International Food Policy Research Institute, Washington, DC, 2002) (www.ifpri.org/2020/books/actionppt/actionppt.pdf).

4. "Making a difference," *World Health Report 1999* (World Health Organization, Geneva, 1999).

5. See, for example, the World Food Programme's goal for 2002–2005: "Excellence in providing food assistance that enables all . . . to survive and maintain healthy nutritional status, and enabling the social and economic development of at least 30 million hungry people every year" (www.wfp.org).

6. United Nations Development Programme, Millennium Development Goals (2003): "The global challenge: Goals and targets" (www.undp.org).

7. E. Lutz, S. Pagiola, C. Reiche, "Economic and Institutional Analysis of Soil Conservation Projects inCentral America and the Caribbean" (Environment Paper No. 8, World Bank, Washington, DC, 1994).

8. *The Daily News*, Harare, Zimbabwe, 17 June 2003 (http://allafrica.com/stories/200306170585.html). The article reports: "Food production in Zimbabwe has fallen by more than 50%, measured against a five-year average, due mostly to the current social, economic and political situation and the effects of the drought," the UN agencies [WFP and FAO] said. The underlying cause was described as "unwise policies over land redistribution."

9. "Global Warming Threatens Food Shortages in Developing Countries" [International Maize and Wheat Improvement Centre (CIMMYT), 2003] (www.cimmyt.cgiar.org/english/webp/support/n_release/warming/nrelease_12May03.htm).

10. FAO Committee on World Food Security, *The Impact of HIV/AIDS on Food Security* (2001) (www.fao.org/docrep/meeting/003/Y0310E.htm).

11. An alternative definition used by the USDA's Natural Resources Conservation Service is, "SOIL QUALITY is how well soil does what we want it to do" (http://soils.usda.gov/sqi/soil_quality/what_is/index.html). The NRCS stresses that soil has both inherent and dynamic properties.

12. J. W. Doran, A. J. Jones, Eds., *Methods for Assessing Soil Quality*. SSSA Special Publication 49 (Soil Science Society of America, Madison, WI, 1996).

13. M. J. Swift, *Nat. Resour.* 35 (no. 4), 12 (1999).

14. P. Sanchez, C. A. Palm, S. W. Buol, *Geoderma* 114, 157 (2003).

15. M. J. Mausbach, C. A. Seybold, in *Soil Quality and Agricultural Sustainability*, R. Lal, Ed. (Ann Arbor Press, Chelsea, MI, 1998), pp. 33–43.

16. C. A. Seybold, M. J. Mausbach, D. L. Karlen, H. H. Rogers, in *Advances in Soil Science*, R. Lal, J. M. Kimble, R. F. Follet, B. A. Steward, Eds. (CRC Press, Boca Raton, FL, 1998), pp. 387–404.

17. A. M. Whitbread, G. J. Blair, R. D. B. Lefroy, *Soil Tillage Res.* 54, 63 (2000).

18. P. Dorward, M. Galpin, D. Shepherd, *Agric. Syst.* 75, 97 (2003).

19. See UN/FAO's Soil Biodiversity Portal on how soil biodiversity can be assessed with the aim of promoting more sustainable agriculture (www.fao.org/ag/AGL/agll/soilbiod/default.htm).

20. E. A. Gyasi, in *Cultivating Biodiversity: Understanding, Analysing and Using Agricultural Diversity*, H. Brookfield, C. Padoch, H. Parsons, M. Stocking, Eds. (ITDG, London, 2002), pp. 245–255.

21. M. Stocking, N. Murnaghan, *Handbook for the Field Assessment of Land Degradation* (Earthscan, London, 2001).

22. F. Kaihura, M. Stocking, *Agricultural Biodiversity in Small-holder Farms of East Africa* (UNU Press, Tokyo, 2003).

23. H. Brookfield, *Exploring Agrodiversity* (Columbia Univ. Press, New York, 2001).

24. P. Sanchez, *Science* 295, 2019 (2002).

25. P. Dreschel, L. A. Gyiele, *The Economic Assessment of Soil Nutrient Depletion* (International Board for Soil Research and Management, Bangkok, 1999).

26. M. Stocking, "Erosion and crop yield," in *Encyclopedia of Soil Science* (Dekker, New York, 2003).

27. M. S. Girvan, J. Bullimore, J. N. Pretty, A. M. Osborn, A. S. Ball, *Appl. Environ. Microbiol.* 69, 1800 (2003).

28. Soil Quality Institute (http://soils.usda.gov/SQI).

29. S. S. Andrews, C. R. Carroll, *Ecol. Appl.* 11, 1573 (2001).

30. Agriculture has a potentially large role in the sequestration of carbon in soils; see (*54*).

31. "Soil science has been brilliantly informed by reductionist physics and chemistry, poorly informed by ecology and geography, and largely uninformed by social science." Quote from M. J. Swift (*55*).

32. D. L. Karlen, C. A. Ditzler, S. S. Andrews, *Geoderma* 114, 145 (2003).

33. J. Boardman, J. Poesen, R. Evans, *Environ. Sci. Policy* 6, 1 (2003).

34. G. Prain, S. Fujisaka, M. D. Warren, Eds., *Biological and Cultural Diversity: The Role of Indigenous Agricultural Experimentation in Development* (Intermediate Technology, London, 1999).

35. P. Crosson, *Soil Erosion and Its On-farm Productivity Consequences: What Do We Know?* (Resources for the Future, Washington, DC, 1995).

36. S. J. Scherr, S. Yadav, *Food Agric. Environ. Disc. Pap. 14* (International Food Policy Research Institute, Washington, DC, 1995).

37. M. Stocking, *Soil Technol.* 1, 289 (1988).

38. A. Tengberg, M. Stocking, S. C. F. Dechen, *Adv. Geoecol.* 31, 355 (1998).

39. M. Stocking, F. Obando, A. Tengberg, in *Sustainable Use and Management of Soils in Arid and Semiarid Regions*, A. Faz, R. Ortiz, A. R. Mermut, Eds. (Quaderna, Murcia, Spain, 2002), vol. 1, pp. 178–192.

40. R. Lal, Ed., *Soil Erosion Research Methods* (Soil and Water Conservation Society, Ankeny, IA, ed. 2, 1994).

41. A. Tengberg, M. Stocking, in *Response to Land Degradation*, E. M. Bridges *et al.*, Eds. (Oxford & IBH, New Delhi, 2001), pp. 171–185.

42. A. Tengberg, M. da Veiga, S. C. F. Dechen, M. Stocking, *Exp. Agric.* 34, 55 (1998).

43. See the USDA-NRCS Web site World Soil Resources for links to most global soil resource sites and commentaries on soil management (www.nrcs.usda.gov/technical/worldsoils).

44. IFPRI 2020 Vision Policy Brief No. 58 (1999) (www.ifpri.org/2020/briefs/number58.htm).

45. Board of the Millennium Ecosystem Assessment, *Living Beyond Our Means, Natural Assets and Human Well-Being—Statement from the Board* (2005) (www.millenniumassessment.org/en/products.aspx).

46. R. Lal, *Science* 304, 1623 (2004).

47. J. N. Pretty, H. Ward, *World Dev.* 29, 209 (2001).

48. For further discussion on the capital assets framework and sustainable natural resource management, the Livelihoods Connect Web site (www.livelihoods.org) contains guidance sheets and a "sustainable livelihoods toolbox."

49. W. Hiemstra, C. Reijntjes, E. van der Werf, Eds., *Let Farmers Judge: Experiences in Assessing the Sustainability of Agriculture* (ILEIA Readings in Sustainable Agriculture, Intermediate Technology, London, 1992).

50. V. Mazzucato, D. Niemeijer, *Rethinking Soil and Water Conservation in a Changing Society* (Tropical Resource Management Papers 32, Wageningen University, Department of Environmental Sciences, 2000).

51. M. N. Versteeg, F. Amadji, A. Eteka, A. Gogan, V. Koudokpon, *Agric. Syst.* 56, 269 (1998).

52. R. Kiome, M. Stocking, *Global Environ. Change* 5, 281 (1995).

53. J. Pretty, *Regenerating Agriculture: Politics and Practice for Sustainability and Self-Reliance* (Earthscan, London, 1995).

54. A. Renwick, A. S. Ball, J. N. Pretty, *Philos. Trans. R. Soc. London Ser. A* 360, 1721 (2002).

Global Freshwater Resources

Soft-Path Solutions for the 21st Century

PETER H. GLEICK

W ater policies in the 20th century relied on the construction of massive infrastructure in the form of dams, aqueducts, pipelines, and complex centralized treatment plants to meet human demands. These facilities brought tremendous benefits to billions of people, but they also had serious and often unanticipated social, economical, and ecological costs. Many unsolved water problems remain, and past approaches no longer seem sufficient. A transition is under way to a "soft path" that complements centralized physical infrastructure with lower-cost community-scale systems, decentralized and open decision-making, water markets and equitable pricing, application of efficient technology, and environmental protection.

The world is in the midst of a major transi-

This article first appeared in *Science* (28 November 2003: Vol. 302, no. 5650). It has been revised and updated for this edition.

tion in water resource development, management, and use. This transition is long overdue. The construction of massive infrastructure in the form of dams, aqueducts, pipelines, and complex centralized treatment plants dominated the 20th-century water agenda. This "hard path" approach brought tremendous benefits to billions of people, reduced the incidence of water-related diseases, expanded the generation of hydropower and irrigated agriculture, and moderated the risks of devastating floods and droughts.

But the hard path also had substantial—often unanticipated—social, economic, and environmental costs. Tens of millions of people have been displaced from their homes by water projects over the past century, including more than 1 million whose villages are now being flooded by the reservoir behind the Three Gorges Dam in China (1). Twenty-seven percent of all North American freshwater fauna populations are now considered threatened with extinction (2), a trend mirrored elsewhere around the world. Adequate

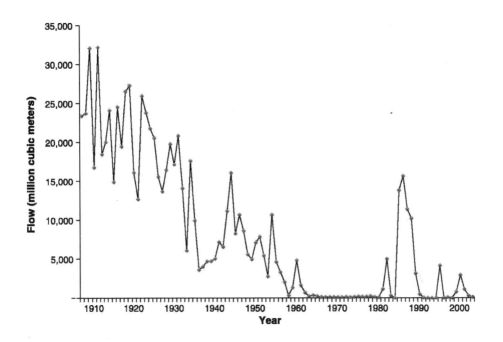

FIGURE 8. Colorado River flows below all major dams and diversions, 1905 to 2001. Data are flows of the Colorado River as measured at U.S. Geological Survey Gage 09-5222, 35 km downstream from Morelos Dam. As shown, flows reaching the Colorado River delta have dropped to near zero in most years.

flows no longer reach the deltas of many rivers in average years, including the Nile, Huang He (Yellow), Amu Darya and Syr Darya, and Colorado, leading to nutrient depletion, loss of habitat for native fisheries, plummeting populations of birds, shoreline erosion, and adverse effects on local communities (3–5).

In arid regions of North America, the hard path for water was pursued especially aggressively. Massive dams and thousands of kilometers of aqueducts were built, permitting humans to withdraw much of the water formerly flowing to wetlands, deltas, and inland sinks, and permitting hydrologic mastery over many watersheds. Since 1905, flows in the Colorado River have decreased markedly because seven states and Mexico withdraw the river's entire flow for agricultural and urban uses (Figure 8). In most years no runoff reaches the river's delta in the Sea of Cortez. Yet calls for new dams in the western United States continue, predicated on the assump-

tion that water problems there will finally be resolved with the construction of another increment of infrastructure (6).

KEY TERM

The phrase **water-related deaths** refers to deaths caused by pathogens or parasites carried in unsanitary drinking water. These could account for more fatalities than the AIDS virus by the year 2020. At present, cholera, typhoid, cryptosporidiosis, *E. coli* infections, and a host of other parasitical, bacterial, and viral-caused illnesses claim more than 5 million lives every year.

Physical Resources

The most serious unresolved water problem is the continued failure to meet basic human needs for water. More than 1 billion people worldwide lack access to safe drinking water; 2.4 billion—more people than lived on the planet in 1940—lack access to adequate sanitation services (7, 8). The failure to satisfy basic water needs leads to hundreds of millions of cases of water-related diseases and 2 million to 5 million deaths annually (9).

Growing awareness of these and other complex challenges led the United Nations General Assembly to declare 2003 the International Year of Freshwater (10). Among the Millennium Development Goals adopted by the General Assembly are new efforts aimed at reducing by half the proportion of people unable to reach or afford safe drinking water and adequate sanitation services by 2015 (11). Without these goals, cumulative water-related deaths by 2020 are expected to be between 52 and 118 million deaths, mostly of children. Even if the water targets are reached, cumulative deaths by 2020 will be between 34 and 76 million. In comparison, cumulative early deaths from AIDS over the same period are projected to be 68 million (12, 13).

Unfortunately, the UN's water goals and solutions to other water problems are unlikely to be achieved, given current levels of financial and political commitments. Despite growing awareness of water issues, international economic support for water projects of all kinds is marginal and declining. Official development assistance for water supply and sanitation projects from countries of the Organisation for Economic Co-operation and Development (OECD) and the major international financial institutions has actually declined over the past few years (Table 1), from $3.4 billion per year (average from 1996 to 1998) to $3.0 billion per year (average from 1999 to 2001). Moreover, those most in need receive the smallest amount of aid. Ten countries received about half of all water-related aid, while countries where less than 60 percent of the population has access to an improved water source received only 12 percent of the money (14).

New challenges further complicate approaches

TABLE 1. Aid to water supply and sanitation by donor (1996–2001). Individual country contributions for water projects in Iraq and Afghanistan have increased in recent years, but overall trends for water-related assistance are unchanged.

Country or multilateral aid organization	Millions of U.S. dollars	
	1996–1998 average	1999–2001 average
Australia	23	40
Austria	34	46
Belgium	12	13
Canada	23	22
Denmark	103	73
Finland	18	12
France	259	148
Germany	435	318
Ireland	6	7
Italy	35	29
Japan	1442	999
Luxembourg	2	8
Netherlands	103	75
New Zealand	1	1
Norway	16	32
Portugal	0	5
Spain	23	60
Sweden	43	35
Switzerland	25	25
United Kingdom	116	165
United States	186	252
Subtotal, countries	*2906*	*2368*
African Development Fund	56	64
Asian Development Bank	150	88
European Community	—	216
International Development Association	323	331
Inter-American Development Bank, Special Operations Fund	46	32
Subtotal, multilateral organizations	*575*	*730*
Total water supply/sanitation aid	3482	3098

to solving water problems. Issues such as regional and international water conflicts (15), the dependence of many regions on unsustainable groundwater use (16), the growing threat of anthropogenic climate change (17), and our declining capacity to monitor critical aspects of the global water balance (18) are all inadequately addressed by water planners and policymakers. If these challenges are to be met within ecological, financial, and social constraints, new approaches are needed.

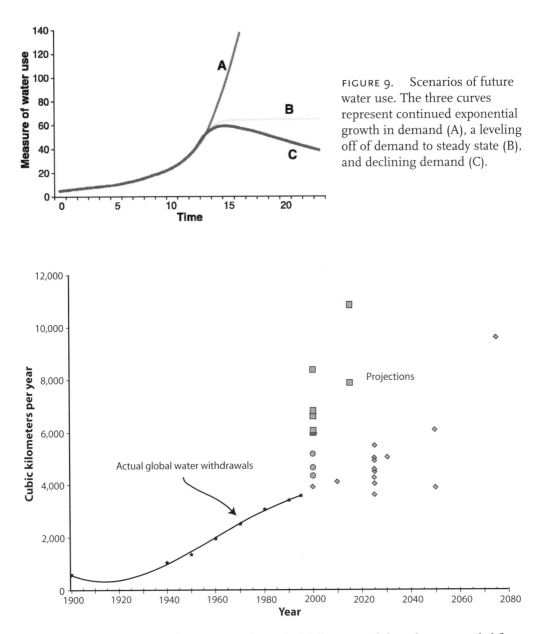

FIGURE 9. Scenarios of future water use. The three curves represent continued exponential growth in demand (A), a leveling off of demand to steady state (B), and declining demand (C).

FIGURE 10. Projections of water use and actual global water withdrawals, as compiled from various projections of global water withdrawals made since the 1960s (44), together with an estimate of actual global water withdrawals, as estimated in (45). Note that projections made before 1980 forecast very substantial increases in water use; more recent forecasts have begun to incorporate possible improvements in water productivity to reflect recent historical experience.

KEY: *squares:* projections made before 1980 (includes forecasts for 2000 or 2015);
circles: projections made between 1980 and 1995 (includes forecasts for 2000);
diamonds: projections made after 1995 (includes forecasts for 2000, 2010, 2025, 2030, 2050, and 2075).

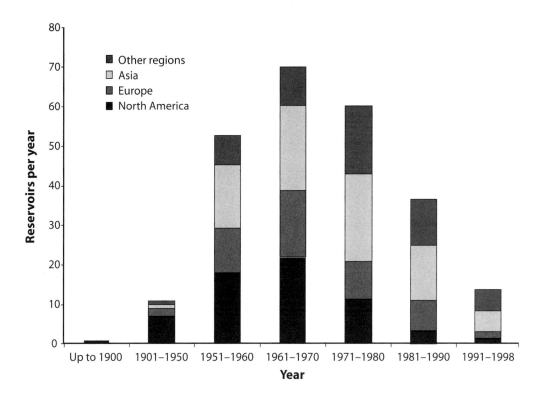

FIGURE 11.   Construction of large reservoirs worldwide in the 20th century. Average numbers of reservoirs with volume greater than 0.1 km³ built by decade, through the late 1990s, are normalized to dams per year for different periods. Note that there was a peak in construction activities in the middle of the 20th century, tapering off toward the end of the century. The period 1991 to 1998 is not a complete decade; note also that the period 1901 to 1950 is half a century. "Other regions" includes Latin America, Africa, and Oceania (46).

20th-Century Water Policy and Planning

The predominant focus of water planners and managers has been identifying and meeting growing human demands for water. Their principal tools have been long-range demand projections and the construction of tens of thousands of large facilities for storing, moving, and treating water. The long construction times and high capital costs of water infrastructure require that planners try to make long-term forecasts and projections of demand. Yet these are fraught with uncertainty.

Three basic futures are possible: (i) exponential growth in water demand as populations and economies grow, (ii) a slowing of demand growth until it reaches a steady state, and (iii) slowing and ultimately a reversal of demand (Figure 9). Reviewing the last several decades of projections shows that planners consistently assumed continued, even accelerated, exponential growth in total water demand (Figure 10). Some projections suggest that water withdrawals would have to triple and even quadruple in coming years, requiring additional dams and diversions of previously untapped water resources in remote or pristine areas once declared off limits to development. Proposals have been made to flood the Grand Canyon, dam the Amazon, and divert Siberian and Alaskan rivers to southern population centers.

Instead, as Figures 10 and 11 show, total water withdrawals began to stabilize in the 1970s and

1980s, and construction activities began to slow as the unquantified but real environmental and social costs of dams began to be recognized. More recently, the economic costs of the traditional hard path have also risen to levels that society now seems unwilling or unable to bear. The most-cited estimate of the cost of meeting future infrastructure needs for water is $180 billion per year to 2025 for water supply, sanitation, wastewater treatment, agriculture, and environmental protection—a daunting figure, given current levels of spending on water (19). This estimate is based on the assumption that future global demand for water and water-related services will reach the level of industrialized nations, and that centralized and expensive water supply and treatment infrastructure will have to provide it. If we focus on meeting basic human needs for water for all with appropriate-scale technology, the cost instead could be in the range of $10 billion to $25 billion per year for the next two decades—a far more achievable level of investment (20). Similarly, as solutions that require large infrastructure investment and environmental disruption have become less attractive, new ideas are being developed and tried and some old ideas—such as rainwater harvesting and integrated land and water management—are being revived. These alternative approaches must be woven together to offer a comprehensive toolbox of possible solutions.

A New Approach for Water

What is required is a "soft path," one that continues to rely on carefully planned and managed centralized infrastructure but complements it with small-scale decentralized facilities. The soft path for water strives to improve the productivity of water use rather than seek endless sources of new supply. It delivers water services and qualities matched to users' needs, rather than just delivering quantities of water. It applies economic tools such as markets and pricing, but with the goal of encouraging efficient use, equitable distribution of the resource, and sustainable system operation over time. And it includes local communities in decisions about water management, allocation, and use (21–23). As Lovins noted for the energy industry, the industrial dynamics of this approach are very different, the technical risks are smaller, and the dollars risked far fewer than those of the hard path (24).

Rethinking water use means reevaluating the objectives of using water. Hard-path planners erroneously equate the idea of using less water, or failing to use much more water, with a loss of well-being. This is a fallacy. Soft-path planners believe that people want to satisfy demands for goods and services, such as food, fiber, and waste disposal, and may not care how much water is used—or even whether water is used at all—as long as these services are produced in convenient, cost-effective, and socially acceptable ways. Thus, society's goal should not be to increase the use of water, but to improve social and individual well-being per unit water used.

Waste disposal, for example, does not require any water, although some water for this purpose may be appropriate or desirable for social or cultural reasons. Industrial nations have grown accustomed to water-based sanitation; indeed, in the United States, the largest indoor user of water in homes is the flush toilet, which requires and then contaminates huge volumes of potable water. In the last two decades, however, the amount of water required by toilets in the United States has declined by up to 75 percent as new efficiency standards have been adopted (25); even greater reductions are possible (26).

Another example of one piece of the soft path can be seen in a new study of the potential for water conservation and efficiency improvements in California's urban sector. This analysis shows that the same services now being provided in urban areas, including residential, commercial, and industrial activities, can be provided with 67 percent of the water now used for those purposes with current technology at current prices (27).

In the agricultural sector, farmers do not want to use water per se; they want to grow crops profitably. Changing irrigation technology or crop characteristics permits growers to produce more

SCIENCE IN THE NEWS

Science, Vol. 303, no. 5661, 1124–1127, 20 February 2004

AS THE WEST GOES DRY
Robert F. Service

MOUNT BACHELOR, OREGON—Under the dome of a concrete-gray sky, Stan Fox assembles four pieces of aluminum tubing into a 3-meter-long hollow pipe. After standing it on end, he plunges it through more than 2 meters of snow at Dutchman Flat, an alpine meadow perched on the shoulder of this 3000-meter mountain. Fox, who heads the Oregon snow-survey program for the U.S. Department of Agriculture's Natural Resources Conservation Service (NRCS), removes the tube and reads the snowpack depth, a measurement that has been tracked at nearby sites monthly since the 1930s. Today the snow is 250 centimeters deep, and by comparing the weights of the tube both filled and empty, Fox and a colleague determine that the snow contains about 30 percent liquid water. If all the snow were instantly liquefied, the water would be nearly 1 meter deep. Not too bad. In a region prone to spikes in precipitation, Dutchman Flat is more than 15 percent above its 30-year average. "The snow in these mountains is a virtual reservoir," Fox says. As the

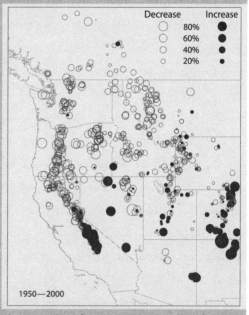

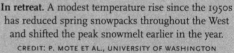

In retreat. A modest temperature rise since the 1950s has reduced spring snowpacks throughout the West and shifted the peak snowmelt earlier in the year.

CREDIT: P. MOTE ET AL., UNIVERSITY OF WASHINGTON

snow melts in the spring and summer, it will slowly release that water, filling streams and reservoirs, which provide lifeblood to the region during the normally bone-dry summer months.

But indications are that this age-old cycle is beginning to change. New assessments of decades' worth of snowpack measurements show that snowpack levels have dropped considerably throughout the American West in response to a 0.8°C warming since the 1950s. Even more sobering, new studies reveal that if even the most moderate regional warming predictions over the next 50 years come true, this will reduce western snowpacks by up to 60 percent in some regions, such as the Cascade Mountains of Oregon and Washington. That in turn is expected to reduce summertime stream flows by 20 percent to 50 percent. "Snow is our water storage in the West," says Philip Mote, a climatologist at the University of Washington (UW), Seattle, who leads a team that has produced much of the new work. "When you remove that much storage, there is simply no way to make up for it."

The impacts could be profound. In the parched summer months, less water will likely be available for everything from agriculture and hydropower production to sustaining fish habitats. Combined with rising temperatures, the dwindling summertime water could also spell a sharp increase in catastrophic fires in forests throughout the West. With much of the current precipitation headed downstream earlier in the winter and spring, the change is also likely to exacerbate the risk of floods.

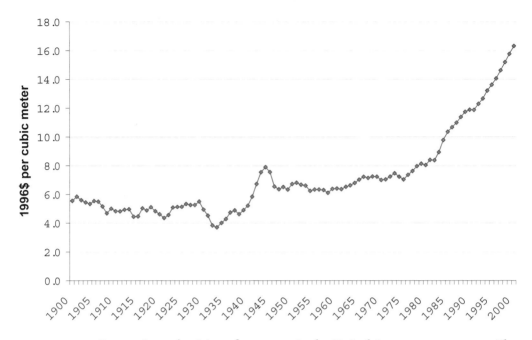

FIGURE 12.   Economic productivity of water use in the United States, 1900 to 2000. The economic productivity of water use in the United States, measured as $GNP (gross national product, corrected for inflation) per cubic meter of water withdrawn, has risen sharply in recent years, from around $6–8/m³ to around $14/m³. Although GNP is an imperfect measure of economic well-being, it provides a consistent way to begin to evaluate the economic productivity of water use.

food and fiber per unit water. Efficiency improvements can result from **furrow diking**, land leveling, direct seeding, drip irrigation, changes in plant varieties, low-energy precision-application sprinklers, and simply better information about when and where to irrigate. Drip irrigation and microsprinklers can achieve efficiencies in excess of 95 percent, compared with flood irrigation efficiencies of 60 percent or less (*28*). As of 2000,

KEY TERM

A technique to capture runoff, **furrow diking** consists of forming soil mounds around planted crops, with troughs or moats in between to hold excess water.

however, the area under micro-irrigation worldwide was less than 3 million hectares, or only about 1 percent of all irrigated land (*29*). In China, furrow and flood irrigation was used on 97 percent of irrigated land; only 3 percent was watered with sprinklers and drip systems (*30*).

Industrial facilities are also finding new ways to reduce water use and recycle existing withdrawals. The dairy industry in the 1970s required 3 to 6 liters of process water to make a liter of milk; today, the most efficient dairies use less than 1 liter of water per liter of milk (*31*). Producing a 200-mm semiconductor silicon wafer used about 30 gallons of water per square inch in 1997, and this is expected to drop to 6 gallons per square inch by the end of 2003 (*32*).

All of these factors have already begun to lead to substantial improvements in the economic productivity of water use, measured in dollars of economic production per cubic meter of water withdrawn. Figure 12 shows the substantial increase in U.S. economic productivity of water from 1900 to 2000 as water-use efficiency has improved and as the U.S. economy has shifted from water-intensive production. Other countries show similar gains (33). These improvements mean that the United States used less water for all purposes in 2000 than it did two decades earlier, despite large increases in our economy and population (34).

Ultimately, meeting basic human and ecological needs for water, improving water quality, eliminating overdraft of groundwater, and reducing the risks of political conflict over shared water require fundamental changes in water management and use. More money and effort should be devoted to providing safe water and sanitation services to those without them, using technologies and policies appropriate to the scale of the problem. Economic tools should be used to encourage efficient use of water and reallocation of water among different users. Ecological water needs should be quantified and guaranteed by local or national laws. And long-term water planning must include all stakeholders, not just those traditionally trained in engineering and hydrologic sciences.

The transition to a comprehensive "soft path" is already under way, but we must move more quickly to address serious unresolved water problems. We cannot follow both paths.

References and Notes

1. World Commission on Dams, *Dams and Development: A New Framework for Decision-Making* (Earthscan, London, 2000).

2. A. Ricciardi, J. B. Rasmussen, *Conserv. Biol.* 13, 1220 (1999).

3. S. Nixon, *Ambio* 23, 30 (2003).

4. C. J. Vorosmarty, M. Meybeck, in *Vegetation, Water, Humans and the Climate*, P. Kabat *et al.*, Eds. (Springer-Verlag, Heidelberg, Germany, 2003), pp. 408–572.

5. M. J. Cohen, C. Henges-Jeck, *Missing Water: The Uses and Flows of Water in the Colorado River Delta Region* (Pacific Institute for Studies in Development, Environment, and Security, Oakland, CA, 2001) (http://www.pacinst.org/reports/missing_water/)

6. Several bills were introduced in 2003 in the U.S. Congress that call for building new dams, expanding existing dams, or studying new sites for dams in the western United States, including HR 2828, introduced by Rep. Ken Calvert (R-CA), and HR 309, introduced by Rep. Devin Nunes (R-CA).

7. World Health Organization, *Global Water Supply and Sanitation Assessment 2000 Report* (www.who.int/docstore/water_sanitation_health/Globassessment/GlobalTOC.htm).

8. The global population in 1940 was 2.3 billion, as estimated in (35).

9. "Water-related diseases" include diseases that are waterborne (diarrheas, dysenteries, and enteric fevers), water-washed (including infectious skin and eye diseases associated with contaminated water), and water-based (including parasitic diseases such as schistosomiasis and dracunculiasis), and diseases associated with water-based insect vectors (such as malaria, yellow fever, and dengue). The lower death estimates exclude water-based insect vector diseases. Data on deaths vary; the World Health Organization estimates deaths from diarrheal diseases alone at 2 million to 3 million per year (7).

10. Resolution A/RES/550196, "International Year of Freshwater, 2003" adopted by the UN General Assembly, 1 February 2001.

11. Resolution A/55/L.2, "United Nations Millennium Declaration" adopted by the UN General Assembly, 18 September 2000. As of late 2005, progress on the water-related goals has been inadequate. New governmental commitments of money, education, and technology will be needed to meet the goals, but it seems unlikely to be forthcoming at the levels necessary.

12. P. H. Gleick, *Dirty Water: Estimated Deaths from Water-Related Diseases 2000–2020* (Pacific Institute for Studies in Development, Environment, and Security, Oakland, CA, 2002) (www.pacinst.org/reports/water_related_deaths_report.doc).

13. The estimate for cumulative AIDS deaths over this same period is 68 million (36).

14. Organisation for Economic Co-operation and Development, Creditor Reporting System, *Aid Activities in the Water Sector: 1997–2002* (OECD, Development Assistance Committee, Paris, 2002).

15. S. Postel, A. Wolf, *Foreign Policy* (September/October), 60 (2001).

16. J. Burke, M. Moench, *Groundwater and Society: Resources, Tensions, Opportunities* (United Nations, New York, 2000).

17. Intergovernmental Panel on Climate Change, *Climate Change 2001: Impacts, Adaptation, and Vulnerability* (Cambridge Univ. Press, Cambridge, 2001), especially the sections on hydrology and water resources.

18. IAHS Ad Hoc Group on Global Water Data Sets, *Eos* 82 (no. 5), 54 (2001).

19. Costs for major hydropower dams or large-scale water transfers are not included in this already-large number (37).

20. The more expensive estimates assume a cost of around $500 per person—typical of the costs of centralized water systems in developed countries. However, field experience shows that safe and reliable water supply and sanitation services can be provided in urban areas for $35 to $50 per person and in rural areas for less than that when local communities build appropriate-scale technology (38).

21. P. H. Gleick, *Nature* 418, 373 (2002).

22. G. Wolff, P. H. Gleick, in *The World's Water 2002–2003*, P. H. Gleick, Ed. (Island, Washington, DC, 2002), pp. 1–32.

23. D. Brooks, *Another Path Not Taken: A Methodological Exploration of Water Soft Paths for Canada and Elsewhere* (Friends of the Earth, Ottawa, Canada, 2003).

24. Lovins is to be credited with coining the term *soft path* for energy (39).

25. National Energy Policy Act of 1992, Public Law 102-486, 24 October 1992.

26. Dual-flush toilets that use only 60 to 70 percent of the water used by toilets that meet current U.S. standards are widely used in Australia and Japan.

27. P. H. Gleick *et al.*, *Waste Not, Want Not: The Potential for Urban Water Conservation in California* (Pacific Institute for Studies in Development, Environment, and Security, Oakland, CA, 2003) (www.pacinst.org/reports/urban_usage/waste_not_want_not_full_report.pdf).

28. A. L. Vickers, *Handbook of Water Use and Conservation* (WaterPlow, Amherst, MA, 2001).

29. S. Postel, *Pillar of Sand: Can the Irrigation Miracle Last?* (Norton, New York, 1999).

30. L. Jin, W. Young, *Water Policy* 3, 215 (2001).

31. *Water Efficiency Manual* (North Carolina Division of Pollution Prevention and Environmental Assistance, 1998) (www.p2pays.org/ref/01/0069206.pdf).

32. In the semiconductor industry, gallons per square inch (gal/in^2) is the standard metric of measuring water use; 1 gal/in^2 is equal to 0.59 liters/cm^2 (40, 41).

33. P. H. Gleick, *Annu. Rev. Environ. Resour.* 28, 275 (2003).

34. GDP data used with permission from L. Johnston, S. H. Williamson, *The Annual Real and Nominal GDP for the United States, 1789–Present. Economic History Services*, April 2002 (www.eh.net/hmit/gdp/). Water data from S. S. Hutson, N. L. Barber, J. F. Kenny, K. S. Linsey, D. S. Lumia, M. A. Maupin, *Estimated Use of Water in the United States in 2000* (USGS Circular 1268, U.S. Geological Survey, 2004. See (42) for data from the 1990s.

35. *World Population Prospects as Assessed in 1963* (United Nations, New York, 1966).

36. UNAIDS, *Report on the Global HIV/AIDS Epidemic*, UNAIDS/02.26.E (Geneva, Switzerland, 2002).

37. W. Cosgrove, F. Rijsberman, *A Water Secure World: Report of the World Commission for Water in the 21st Century*. Chapter 4: Framework for Action (World Water Council, London, 2000) (www.worldwatercouncil.org/Vision/Documents/CommissionReport.pdf).

38. Water Supply and Sanitation Collaborative Council, *Vision 21: A Shared Vision for Hygiene, Sanitation, and Water Supply* (2000) (www.worldwatercouncil.org/Vision/Documents/VISION21FinalDraft.PDF).

39. A. Lovins, *Soft Energy Paths: Toward a Durable Peace* (Ballinger, Cambridge, MA, 1977).

40. *International Technology Roadmap for Semiconductors, 2001 Edition* (http://public.itrs.net/Files/2001ITRS/Home.htm).

41. S. Allen, M. R. Hahn, *Semiconductor Wastewater Treatment and Reuse. Semiconductor Fab Tech* (Microbar Inc., Sunnyvale, CA, ed. 9, 1999).

42. W. B. Solley, R. R. Pierce, H. A. Perlman, *Estimated Use of Water in the United States, 1995* (USGS Circular 1200, U.S. Geological Survey, Denver, CO, 1998).

43. P. H. Gleick, in *The World's Water 2000–2001*, P. H. Gleick, Ed. (Island, Washington, DC, 2000), pp. 39–61.

44. For a summary of global water scenarios and a complete list of the scenarios and projections used in this figure, see (43).

45. I. A. Shiklomanov, *Assessment of Water Resources and Water Availability in the World. United Nations Report for the Comprehensive Assessment of the Freshwater Resources of the World*, Data archive, CD-ROM from the State Hydrologic Institute of St. Petersburg, Russia (1998).

46. A. B. Avakyan, V. B. Iakovleva, *Lakes Reservoirs Res. Manage.* 3, 45 (1998).

Web Resources

www.sciencemag.org/cgi/content/full/302/5650/1524/DC1

Energy Resources
and Global Development

JEFFREY CHOW, RAYMOND J. KOPP,
and PAUL R. PORTNEY

I n order to address the economic and environmental consequences of our global energy system, we consider the availability and consumption of energy resources. Problems arise from our dependence on combustible fuels, the environmental risks associated with their extraction, and the environmental damage caused by their emissions. Yet no primary energy source, be it renewable or nonrenewable, is free of environmental or economic limitations. As developed and developing economies continue to grow, conversion to and adoption of environmentally benign energy technology will depend on political and economic realities.

Energy is the lifeblood of technological and economic development. The energy choices made by the United States and the rest of the world have

This article first appeared in *Science* (28 November 2003: Vol. 302, no. 5650). It has been revised and updated for this edition. The authors thank J. Darmstadter and R. Newell for reviewing and providing comments on this paper.

ramifications for economic growth; for the local, national, and global environments; and even for the shape of international political alliances and national defense commitments. Countries of varying levels of wealth also face different energy challenges (1). Here we discuss the availability of global energy resources, how they are used and by whom, and the consequences of the global distribution and use of energy resources.

Although estimates vary, the world's proved, economically recoverable fossil fuel reserves include almost 1 trillion metric tons of coal, more than 1 trillion barrels of petroleum, and over 150 trillion cubic meters of natural gas (Table 2) (2). In addition to fossil fuels, mineral resources important to energy generation include over 3 million metric tons of uranium reserves (3). To put this into context, consider that the world's annual 2000 consumption of coal was about 5 billion metric tons or 0.5 percent of reserves. Natural gas consumption was 1.6 percent of reserves, whereas oil was almost 3 percent of reserves, and nuclear electricity generation consumed the

TABLE 2.   Proved reserves of mineral energy resources, their approximate energy content, and their distribution among income groups in 2001. Petroleum, coal, and natural gas estimates are calculated from data in (28). Uranium estimates are calculated from data in (29).

Income group	Petroleum			Coal			Natural gas			Uranium		
	10^9 barrels	EJ*	% total	10^6 metric tons	EJ	% total	10^{12} cubic m	EJ	% total	1000 metric tons	EJ	% total
Poorest 10%	24	140	2.33	475.4	10	0.05	3.7	140	2.41	193.5	15000	5.91
Low–income	49.5	290	4.80	133090.5	2700	13.54	12.9	490	8.37	488.8	38000	14.92
Middle-income	723.6	4300	70.14	439213.1	8800	44.69	105.2	4000	68.12	1282.2	99000	39.14
Low– + middle–income (developing)	773.1	4500	74.93	572303.6	11000	58.24	118.2	4500	76.49	1771	140000	54.06
High–income (developed)	258.6	1500	25.07	410409.1	8200	41.76	36.3	1400	23.51	1490.4	110000	45.49
Richest 10%	38.6	230	3.74	317104.1	6300	32.27	9.5	360	6.13	332.1	26000	10.14
Total	1031.7	6100		982712.7	20000		154.5	5900		3276.1	250000	

* EJ = exajoules = 10^{18} joules (9.48 x 10^{14} Btus).
Proved reserves of uranium refer to RARs described by the International Atomic Energy Agency.

equivalent of 2 percent of uranium reserves (4). Reported recoverable reserves have tended to increase over time, keeping pace with consumption, and now are at or near all-time highs. In relation to current consumption, there remain vast reserves that are adequate for continued worldwide economic development, not even accounting for reserves that will become economically recoverable through continuing discovery and technological advance. Thus, it seems that the world is not running out of mineral fuels.

It should be noted that the three major fossil fuels are not perfect substitutes for each other, particularly in the short term. Petroleum derivatives offer versatility in use and ease of transport that make them ideal for the transportation sector. Coal is the most abundant fossil fuel but generates the most airborne pollutants. Hence, coal-fired electricity generation plants are gradually giving way to gas-fired plants. Natural gas is the cleanest-burning and most energy-efficient fossil fuel, but supply is currently hindered by insufficient extraction and transport infrastructure, such as regasification and storage facilities for importing liquefied natural gas from overseas.

Large fossil fuel reserves are concentrated in a small number of countries, with half of the low-income countries and more than a third of the middle-income countries having no fossil fuel reserves whatsoever. Similarly, the majority of reserves in the developed countries also are concentrated in a relatively few nations, notably the United States and several of the wealthier oil-producing Middle Eastern states (5). If energy reserves were necessary for economic development, several of the world's poorest nations would be disadvantaged. However, many energy-bereft countries (such as Japan) have become highly developed through sufficient access to international energy markets.

Conversely, Nigeria possesses substantial reserves but remains one of the poorest countries, its energy production activities mired in corruption. Thus, simply possessing large fossil energy reserves is of questionable value to a country's development if there is no well-functioning and adequately equitable socioeconomic system enabling it to extract and deploy those energy resources for their full social benefit.

Total global energy use exceeds 370 **exajoules** (EJ) [350 quadrillion **British thermal units (Btus)**] per year, which is equivalent to over 170 million barrels of oil each day (6).

Approximately 95 percent of this energy comes from fossil fuels. Global energy consumption draws from six primary sources: 44 percent petro-

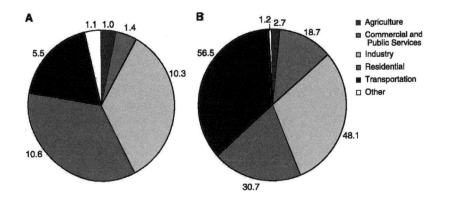

FIGURE 13. Per-capita energy consumption by sectoral end use in
(A) the developing world and (B) the developed world (in gigajoules)

leum, 26 percent natural gas, 25 percent coal, 2.5 percent hydroelectric power, 2.4 percent nuclear power, and 0.2 percent nonhydro renewable energy (7, 8). A considerable amount of primary energy is converted to electricity either in the course of initial harvesting (as for hydroelectric, wind, and geothermal) or by combustion (as for fossil, biomass, and waste fuels). These estimates do not include nonmarket fuelwood and farm residues that are prevalent in many developing countries, because global estimates of noncommercial energy use are often incomplete and unreliable. However, the International Energy Agency (IEA) suggests that biomass provides, on average, one-third of the energy needs in Africa, Asia, and Latin America, and as much as 80 to

90 percent in the poorest countries of these regions (9).

Processing and conversion of primary sources permit enormous versatility in energy use in many different settings. The end applications of this consumption can be categorized into five major sectors: industry, transportation, agriculture, commercial and public services, and residential (10). Developing countries use the most energy in the residential sector (11), followed by industrial uses and then transportation (Figure 13A). The opposite is true for developed countries, where transportation consumes the largest amount of energy, followed by industrial and then residential consumption (Figure 13B).

Unsurprisingly, the developing and industrialized worlds demonstrate striking disparities in annual energy consumption per capita (12). Industrialized country energy use exceeds that of the developing countries for all five end-use sectors by 3 to 14 times, depending on the sector (Figure 13, A and B) (13). In aggregate, the average person in the developing countries consumes the equivalent of 6 barrels of oil [34 gigajoules (GJ) or 32 million Btus] annually, whereas the average person in the developed world consumes nearly 40 barrels (220 GJ or 210 million Btus) (Table 3) (14). Residents of the poorest 10 percent of countries consume less than one barrel of oil equivalent per year per capita, whereas their counterparts in the richest 10 percent of countries consume over 60

TABLE 3.   Aggregate energy consumption per capita and per US$ GDP in 2000.

Income Group	per capita		per US$ GDP	
	10^6 BTUs	gigajoules	10^3 BTUs	megajoules
Poorest 10%	3.4	3.6	15.0	15.8
Low income	12.7	13.4	28.9	30.5
Middle income	49.9	52.6	23.9	25.2
Low + middle income (developing)	32.1	33.9	24.6	26.0
High income (developed)	211.7	223.4	7.5	7.9
Richest 10%	218.0	230.0	7.0	7.4

Although it appears that individuals in wealthier countries consume substantially more energy than those in developing countries, wealthier countries are actually more efficient in terms of the energy intensity of Gross Domestic Product (GDP).

times as much. Also striking are disparities within the developing world, as well as the fact that, on an annual per-capita basis, the middle-income countries use four times as much energy as their low-income counterparts. These relationships between wealth and energy consumption suggest that as a country becomes richer, its people tend to consume substantially more energy (Table 3).

However, looking at energy use within the high-income group alone, the correlation is weaker. For example, Norway has a gross national income per capita (GNI/pop) of US$34,530; Japan is slightly higher at US$35,620, but energy consumption per capita is lower in Japan: 150 GJ compared to 250 GJ for Norway. This discrepancy is probably due to the availability of relatively inexpensive hydroelectric power in Norway, whereas Japan, possessing fewer local resources, has greater incentives to be more energy efficient. Therefore, although at a coarse scale energy consumption per capita increases with economic growth, there are different paths that a particular country's energy system can take in its development, with some paths resulting in greater efficiency and less consumption than others.

Moreover, when one examines energy use per dollar of gross domestic product (GDP), low-income countries use more energy to create a dollar of GDP than do high-income countries, because of greater use of more energy-efficient technologies as a country develops (Table 3) (15). Furthermore, as cleaner energy-efficient technologies generated by the industrialized countries become cheaper, the growing economies of the developing world become more likely to adopt them, bypassing more wasteful and polluting technologies. For example, countries such as Brazil, China, India, and the Philippines have been installing high-voltage direct-current cables to deliver electricity with greater reliability and efficiency than the alternating-current cables prevalent in the United States.

The data also strongly indicate that the world is heavily dependent on fossil fuel energy, with only about 5 percent coming from other sources. It will remain so, barring substantial technological change. In the near term, this continued dependence is partially because of the paucity of convenient alternatives to petroleum products as fuels in the transportation sector, which consumes the most energy in the developed world. Currently, transportation in the poorest decile of countries consumes less than 3 percent of the energy consumed by that sector in the richest countries.

As developing countries become richer and expand their transportation networks, petroleum products will likely fuel them. The industrial sector of the developed world also relies heavily on fossil fuels. Institutional inertia, as well as the cost of replacing capital-intensive, fossil energy–dependent infrastructure, slow the pace at which nonfossil substitutions can occur.

Between 1980 and 2001, worldwide consumption of petroleum, coal, and natural gas increased by 22, 27, and 71 percent, respectively. Concurrently, annual world carbon dioxide (CO_2) emissions from the consumption and flaring of fossil fuels, implicated as the predominant cause

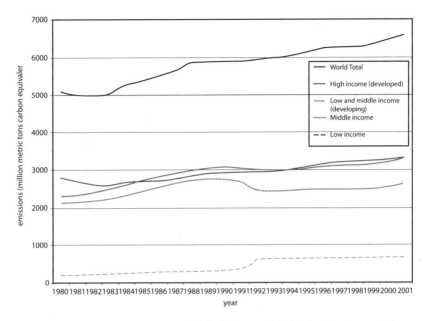

Emissions have tended to increase over time in both the developing and developed worlds. Increases in the industrialized countries can be attributed to increasing per capita emissions, while the increases in the developing world are due to a combination of population growth and growth in per capita emissions. The decline among middle-income countries and increase among low-income countries illustrated in the early 1990s was due to the restructuring of political boundaries and economic downturn following the collapse of the Soviet Union. This figure otherwise categorizes countries by their income group in year 2000, ignoring mobility between groups over the period depicted. Data were obtained from (28).

FIGURE 14. Trends in carbon dioxide emissions due to consumption and flaring of fossil fuels, by income group.

of global climate change, increased from 5 billion to 6.6 billion metric tons carbon equivalent, with relatively steady increases occurring for all income groups (Figure 14). Fossil fuel consumption also results in emissions of other greenhouse gases (GHGs), such as carbon monoxide, methane, and volatile organic compounds (VOCs), not to mention nitrous oxides

---

KEY TERM

The **sequestration capacity** of a natural sink is the amount of CO_2 that can be stored in natural systems that absorb more carbon than they emit, such as oceans and forests.

(NO_x) that facilitate the formation of heat-trapping tropospheric ozone.

Although fossil fuel reserves are in no danger of diminishing in the foreseeable future, should the world continue to consume all or even a large fraction of those resources though normal combustion processes, the release of additional GHGs into the atmosphere would likely have substantial consequences for the global climate. According to the Intergovernmental Panel on Climate Change (IPCC), climate models generally predict that continued emissions of anthropogenic GHGs beyond the **sequestration capacity** of natural sinks will result not only in increased mean temperatures but also in more frequent extreme climate events such as droughts and intense storms, with significant consequences for ecosystems, agricultural productivity, and human welfare (16).

Besides GHG emissions, fossil fuel production and use come with other environmental costs. Fossil fuel exploration requires seismic testing and road building that can harm wildlife and fragment habitats. Extraction requires replacing natural habitat with infrastructure and can lead to leaking of fuels and toxic by-products—such as arsenic, cadmium, and mercury—into the local environment.

In the case of oil, spills that occur during transportation and refining also damage local environments. Sulfur dioxide, NO_x, and VOCs released from coal and petroleum combustion cause a myriad of environmental problems, including acid rain, smog, and nitrogen loading.

Furthermore, the traditional alternatives to fossil energy—hydroelectricity and nuclear power—have environmental and social costs that limit their viability as long-term fossil fuel substitutes. In addition to the drawback that there are few rivers left that are ideal for damming to produce electricity, hydroelectricity infrastructure causes dramatic alterations in riparian ecosystems and often the inundation of human settlements and terrestrial habitat. Fissile nuclear power, too, is unlikely to expand because of objections to waste disposal and concerns over weapons proliferation.

It should be noted that no primary energy source and its associated technology is completely free of environmental and other drawbacks. Wind-powered electric turbines require the installation of infrastructure, can cause the death of migratory birds, and elicit local objections on esthetic grounds. Geothermal plants emit CO_2 and hydrogen sulfide. Wind, solar, and geothermal systems are capital-intensive and their viability is geographically limited (17). Without affordable and practical electricity storage, intermittency is also a problem for wind and solar power. The domestically combusted biomass used in developing countries is often a health hazard because of indoor smoke inhalation (18), and mass-produced fuels derived from biomass place greater burdens on agricultural and forest productivity. Even the highly touted hydrogen **fuel cell**, which releases only water vapor, would initially require fossil fuels as hydrogen fuel stock (19). In order to minimize environmental damage relative to the benefits of energy consumption, a sustainable, environmentally benign energy system, or at least the transition to one, will involve a heterogeneous

TABLE 4. Current average European costs of conventional and renewable energy production (1/100th of a Euro in base year 1990 values per kilowatt-hour). PVs, photovaltaics; 1 Euro = US$1.17.

Conventional energy			Renewable alternatives		
Coal	GTCC	Nuclear	Wind	Solar PVs	Bio-energy
3.4	3.1	4.0	7.1	65.5	4.8

Source: Commission of the European Communities, *Green Paper Towards a European Strategy for the Security of Energy Supply* (Commission of the European Communities, Brussels, 2000).

portfolio of renewable primary sources in order to minimize the environmental impact of any particular source.

The environmental costs of fossil, hydroelectric, and nuclear energy consumption could drive the world toward alternative sources before scarcity becomes a significant issue. Government programs to reduce the negative environmental effects of fossil fuel production and consumption have the same effect as scarcity-induced price increases, and would stimulate (or mandate) new energy technologies that increase efficiency, mitigate pollution, and substitute for fossil energy. Policy mechanisms to achieve these ends include environmental standards, fuel and emission taxes, subsidies for renewable energy production, mandated diversified energy portfolios, and emission **permit–trading schemes**. In the United States and elsewhere, several of these policies (such as regulated limits, emission fees, and tradable permits) have been successfully implemented to reduce noncarbon air pollution, improve air quality, and reduce acid rain (20–23).

Given growing environmental concerns, the future use of fossil resources will likely not follow the standard combustion path of the past but will involve processes with increased efficiency, lower localized air pollution, and perhaps carbon capture and sequestration before, after, or instead of combustion (24). Electricity in particular will remain the most important end-use energy form because of its flexibility in both generation and use. Renewable sources of electricity from solar, wind, geothermal, and tidal power are currently available, but they remain the least consumed form of energy across all income groups (25). Per-capita consumption rates do not exceed 1 MJ (100,000 Btus) per year in the developing-country categories (less than a gallon of oil equivalent), and do not exceed 1 GJ (1 million Btus) per year in the high-income category, with only 24 industrialized countries consuming significant amounts (26).

Renewable energy sources will become prevalent only if they can be more competitive than fossil fuels in relative prices (Table 4). Competition from lower-cost conventional power production, notably by **gas turbine combined cycle** (GTCC) systems, will continue to undercut renewables, even with falling costs (27). Rather than wait for scarcity-induced price rises, governments can accelerate the adoption of renewables with two coordinated and self-reinforcing actions. First, governments can adopt a variety of R&D polices

KEY TERM

In a **gas turbine combined cycle (GTCC) system**, the exhaust from a gas turbine is fed into a residual heat boiler that generates steam for a bottoming steam turbine cycle. If natural gas is used to fuel the gas turbine, the overall efficiency of the system can be slightly more than 50 percent (which compares favorably with the efficiency of traditional heating systems, which are between 10 and 35 percent).

Science, Vol. 305, no. 5686, 967, 13 August 2004

CAN THE DEVELOPING WORLD SKIP PETROLEUM?

By Gretchen Vogel

If technologies for hydrogen fuel take off, one of the biggest winners could be the developing world. Just as cell phones in poor countries have made land lines obsolete before they were installed, hydrogen from renewable sources—in an ideal world—could enable developing countries to leap over the developed world to energy independence. "The opportunity is there for them to become leaders in this area," says Thorsteinn Sigfusson of the University of Iceland, one of the leaders of the International Partnership for a Hydrogen Economy (IPHE), a cooperative effort of 15 countries, including the United States, Iceland, India, China, and Brazil, founded last year to advance hydrogen research and technology development.

Different future. Countries not yet committed to fossil fuels might go straight to hydrogen.
CREDIT: SASCHA BURKARD/ISTOCKPHOTO.COM

With their growing influence in global manufacturing, their technical expertise, and their low labor costs, Sigfusson says, countries such as China and India could play extremely important roles in developing more efficient solar or biotech sources of hydrogen—as well as vehicles and power systems that use the fuel. "They have the opportunity to take a leap into the hydrogen economy without all the troubles of going through combustion and liquid fuel," he says. The impact would be huge. The IPHE members already encompass 85 percent of the world's population, he notes.

The current steps are small. For example, a joint U.S.-Indian project is working to build a hydrogen-powered three-wheel test vehicle. The minicar, designed for crowded urban streets, needs only one-tenth as much storage space as a standard passenger car. India's largest auto manufacturer, Mahindra and Mahindra, has shipped two of its popular gasoline-powered three-wheelers (currently a huge source of urban air pollution), to the Michigan-based company Energy Conversion Devices (ECD). Engineers at ECD are working to convert the engine to run on hydrogen stored in a metal hydride. One model will return to India for testing, and one will remain in the United States. The small project "is just the beginning," says a U.S. Department of Energy official. "But the point of bringing in these countries is that they are huge energy consumers. They simply have to be part of the partnership, especially as we start to use the technologies."

(usually in the form of subsidies) that would bring down the price and improve the performance of renewables in comparison with fossil fuels. Second, they can raise the price of fossil fuels through carbon taxes or permits, and thereby tilt the economics toward renewables. These actions serve to push renewables forward by subsidizing their development, while pulling renewables into the market by disadvantaging the price competitiveness of fossil fuels.

As the major blackouts in North America and Italy in 2003 made clear, even energy systems in the richest countries are far from problem-free. Similar systems in the developing world may be even more fraught with trouble as they develop. However, sub-Saharan Africa and other poor countries will probably never have an electricity grid exactly like those of today's high-income countries, even when they have pulled themselves out of wrenching poverty.

In the same way that the developing world is bypassing the paired-copper-wire grid that characterizes telephony in the developed world, leaping directly to cellular communication, so too is it likely to rely much less heavily on our technological model of electricity generation. Rather than adopting a system with large central station power plants generating electricity and distributing it over long distances, we speculate that the developing countries, especially the poorest, are more likely to eventually adopt smaller and less capital intensive microturbines and renewable sources of electricity generation such as biomass, wind, and solar that are closer to the point of use. These applications will bring with them their own sets of problems, but will enable the developing world to avoid others.

Will the world make a transition to alternative, more renewable sources of energy? The simple answer is yes, if only because, in time, supplies of fossil fuels will become too costly. For the next 25 to 50 years, however, this seems not to be a likely prospect. With energy choices driven by relative prices, fossil fuels will dominate energy use for many years to come. These fuels remain relatively inexpensive, and they are supported by a very broad and long-lived infrastructure of mines, wells, pipelines, refineries, gas stations, power plants, rail lines, tankers, and vehicles. Very powerful political constituencies exist worldwide to ensure that investments in this infrastructure are protected. If fossil fuel depletion occurs more rapidly than we expect, or if governments enact policies that artificially increase fossil fuel prices, renewables and alternative energy sources may come online more quickly. The requisite political will and financial support to enact such changes will occur only when societies and their governments decide that the benefits of fossil fuel consumption do not make up for the negative effects on environmental health and human welfare of fossil fuel dependence.

References and Notes

1. We analyzed year 2000 data from 211 countries, using the World Bank's method of distinguishing between low-, middle-, and high-income countries according to GNI/pop. We refer to low- and middle-income countries jointly as developing countries, and high-income countries are considered industrialized or developed countries. Of the countries considered in this analysis, approximately 75 percent fall into the former category. Countries are low-income if GNI/pop is less than US$750 (69 countries, including the Congo, India, and Indonesia); middle-income if GNI/pop is between US$750 and $9,250 (85 countries, including Argentina, Mexico, and Turkey); or high-income if GNI/pop is greater than US$9,250 (57 countries, including the United States, Japan, and Western Europe). We have also identified those countries comprising the poorest 10 percent (such as Cambodia, Chad, and Tajikistan) and the richest 10 percent (such as Singapore, the United Kingdom, and the United States). The developing-country group is heterogeneous in resource endowments and development conditions, whereas classification as a developed country does not imply a preferred or final stage of development. GNI/pop is a convenient criterion among many metrics for levels of development and does not necessarily reflect development status. GNI, GDP, and population data for 2000 are drawn from the *World Development Indicators 2002*, published by the World Bank. Population, GNI/pop, and income categorization for all 211 countries are available at (30).

2. These numbers are based on year 2001 data from (28). Reserves include only resources that are identified as economically and technically recoverable with current technologies and prices. Other resources with foreseeable or unknown potential for recovery exist but are not included in this report, because estimates are often highly speculative and unreliable, particularly estimates of resources in developing countries. Reserve estimates tend to expand overall with time, as technology increases the number of economically recoverable reserves.

3. These numbers are based on year 2001 data from (29). This estimate includes reasonably assured resources (RARs) identified by the IAEA and does not include other potential resources and secondary supplies from reprocessed uranium, reenriched uranium, and highly enriched uranium from the dismantlement of nuclear weapons. A list of reserves by country is available at (30).

4. However, 42 percent of uranium used for nuclear electricity generation is currently supplied by secondary

sources, so the actual consumption of uranium reserves is less than this estimate suggests.

5. A map and list of global reserves by country as well as a more detailed descriptive analysis are available at (30).

6. These numbers are based on year 2000 data from (28).

7. Renewables include energy generated from sources such as geothermal, wind, solar, wood, and waste fuels. This percentage does not include the domestic use of fuelwood and other biomass common in developing countries, but does include energy derived from electric power generation using these fuels.

8. Global maps and tables of consumption by energy source are available at (30).

9. S. L. D'Apote, in *Biomass Energy: Data, Analysis and Trends* (International Energy Agency, Vienna, 1998), pp. 1–31.

10. The year 2000 data used to examine sectoral uses of energy are drawn from *Energy Balances of the OECD Countries* and *Energy Balances of the Non-OECD Countries*, compiled by the International Energy Agency. These data exist only for 133 countries and are not directly comparable with the data discussed above for 200-plus countries provided by the Energy Information Administration (EIA) of the U.S. Department of Energy. This data set includes the consumption of combustible renewables and waste, such as fuelwood, whereas the EIA data set does not. Inferences drawn from direct comparison of the two different sets of data and analyses would not be robust. More detailed descriptions of end-use sectors as defined by the International Energy Agency are available at (30).

11. Residential energy consumption in many regions that are included among the developing states consists predominantly of combustible materials and waste such as fuelwood, manure, and other biofuels, rather than the forms of energy described in the analyses above. Biomass is often the only available and affordable source of energy for basic needs, such as cooking and heating, for large portions of rural populations and for the poorest sections of urban populations in developing countries.

12. Maps and tables of per-capita aggregate energy consumption by country are available at (30).

13. Country tables of per-capita energy consumption by end use are available at (30).

14. These numbers are based on year 2000 data from (28). Per-capita consumption is calculated by dividing aggregate energy consumption by population and does not account for imports and exports of energy embodied in the trade of goods.

15. It is generally accepted practice to use GDP rather than GNI when discussing the energy intensity of economic output. Our conclusions would be no different if we used GNI. Global maps and tables of aggregate energy consumption per dollar of GDP are available at (30).

16. IPCC, *Climate Change 2001: The Scientific Basis* (IPCC, Cambridge, 2001).

17. Coastal areas and plains are ideal for wind power, sunny areas such as equatorial regions for solar power, and volcanic basins for geothermal energy.

18. These fuels are also often used inefficiently because of poor technology (such as a lack of closed stoves or ventilation) and have negative health effects, depending on their method of use. Thus, the health hazards associated with traditional biomass are partly the consequence of sociocultural and developmental problems and can be mitigated with simple technological improvements.

19. In the case of hydrogen fuel cells, one must also consider the environmental consequences of increased levels of water vapor in the atmosphere, should this technology be widespread in the future.

20. D. Burtraw *et al.*, *Resources for the Future Discussion Paper 97-31* (Resources for the Future, Washington, DC, 1997).

21. A. Blackman, W. Harrington, *Resources for the Future Discussion Paper 98-21* (Resources for the Future, Washington, DC, 1998).

22. A. Krupnick *et al.*, *Resources for the Future Discussion Paper 98-46* (Resources for the Future, Washington, DC, 1998).

23. D. Burtraw, K. Palmer, *Resources for the Future Discussion Paper 03-15* (Resources for the Future, Washington, DC, 2003).

24. S. Anderson, R. Newell, *Resources for the Future Discussion Paper 02-68* (Resources for the Future, Washington, DC, 2003).

25. Renewables referred to in this discussion include electricity generated from geothermal, solar, wind, biomass, and waste sources, but not domestically combusted fuels.

26. These numbers are based on year 2000 data from (28).

27. J. Darmstadter, *Resources for the Future Issue Brief 02-10* (Resources for the Future, Washington, DC, 2002).

28. *International Energy Annual 2001 Edition* (EIA, U.S. Department of Energy, Washington, DC, 2003).

29. *Analysis of Uranium Supply to 2050* (International Atomic Energy Agency, Vienna, 2001).

30. See www.rff.org/energyresources/.

31. *Oil Gas J.* 99, 125 (2001)

Global Air Quality and Pollution

HAJIME AKIMOTO

The impact of global air pollution on climate and the environment is a new focus in atmospheric science. Intercontinental transport and hemispheric air pollution by ozone jeopardize agricultural and natural ecosystems worldwide and have a strong effect on climate. Aerosols, which are spread globally but have a strong regional imbalance, change global climate through their direct and indirect effects on radiative forcing. In the 1990s, nitrogen oxide emissions from Asia surpassed those from North America and Europe and should continue to exceed them for decades. International initiatives to mitigate global air pollution require participation from both developed and developing countries.

When the first measurements of high concentrations of CO over tropical Asia, Africa, and

South America were made available by the MAPS (Measurement of Air Pollution from Satellite) instrument launched in 1981 on the space shuttle *Columbia* (1), it became clear that air pollution was a global issue.

Those images showed not only that industrial air pollution from fossil fuel combustion could affect regional and global air quality, but that emissions from biomass burning (forest fires, agricultural waste burning, and vegetable fuel combustion) were important as well, confirming the hypothesis of Crutzen *et al.* (2). This meant that people in less-developed countries, as well as residents of industrialized and rapidly growing developing countries, could suffer from air pollution generated elsewhere. Another illustration of the global character of air pollution came from measurements of tropospheric ozone made by the TOMS (Total Ozone Mapping Spectrometer) and SAGE (Stratospheric Aerosol and Gas Experiment) instruments on the *Nimbus 7* satellite (3). Once again, the impact of biomass

This article first appeared in *Science* (5 December 2003: Vol. 302, no. 5651). It has been revised and updated for this edition.

Ozone is a colorless gas (O_3) that can be produced by electric discharge in oxygen or by the action of ultraviolet radiation on oxygen in the stratosphere, where it acts as a screen for ultraviolet radiation. It can also act as a dangerous component of smog when found near the Earth's surface.

The layers of the atmosphere are defined by critical points in altitude and temperature. The **troposphere** is the lowest atmospheric layer, closest to the Earth's surface. It ranges from 4 to 11 miles high, depending on the latitude at which it is measured. The destruction of the ozone layer by CFCs, hypothesized by Rowland and Molina in 1974, occurs primarily in the polar troposphere and stratosphere.

The **atmospheric lifetime** of a trace gas is derived from the sum of its sinks, or annual loss rates. If the atmospheric burden of a gas is 100 gigatons, and the mean global sink is currently 10 gigatons per year, the lifetime is 10 years. Tropospheric ozone has an atmospheric lifetime of weeks or months; perfluoroethane has a lifetime of 10,000 years.

burning on regional ozone concentrations was demonstrated, in addition to that of industrial pollution. More recently, observations of various tropospheric air pollutants such as NO_2, SO_2, and HCHO by GOME (Global Ozone Monitoring Experiment) and SCHIAMACHY (Scanning Imaging Absorption Spectro-Meter for Atmospheric ChartographY) (4) and of CO by MOPITT (Measurement of Pollution in the Troposphere) (5) have revealed pollution on a global scale.

Edwards *et al.* (6) obtained a picture of the processes affecting tropospheric O_3 production over Africa and the Atlantic, combining the data on TOMS O_3, MOPITT CO, and GOME NO_2. Aerosols are another category of air pollutants that can be viewed from satellites. This and other studies (7–10) show that satellite data can be useful for revealing climatic and environmental implications of global air pollution.

Global air-quality issues exist only in regard to those pollutants that have atmospheric lifetimes long enough (on the order of 1 week) for them to be transported at least to another continent. One such trace gas is tropospheric ozone, a potent greenhouse gas (11) that also is toxic to humans, animals, and plants. Because the atmospheric lifetime of ozone is 1 to 2 weeks in summer and 1 to 2 months in winter (12), ozone produced in a polluted region of one continent can be transported to another continent all year long.

Hemispheric transport, which typically takes place in about 1 month, can occur in all seasons except summer. Figure15 shows how model-calculated surface O_3 during the growing season (May through August) in the Northern Hemisphere has increased between 1860 and 1993 (13). According to this analysis, the concentration of surface O_3 over the mid- and high-latitude

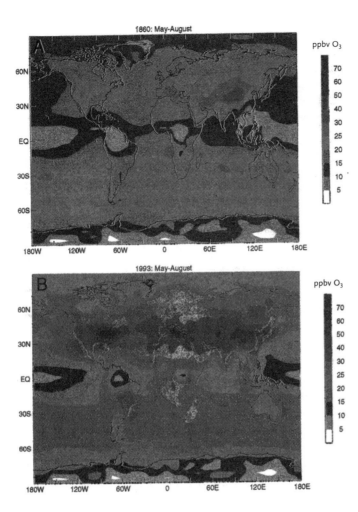

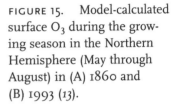

FIGURE 15.   Model-calculated surface O_3 during the growing season in the Northern Hemisphere (May through August) in (A) 1860 and (B) 1993 (13).

Eurasian and North American continents was 15 to 25 parts per billion by volume (ppbv) in 1860 but has increased to 40 to 50 ppbv even in relatively remote areas, and from 10 to 15 ppbv to 20 to 30 ppbv over the mid- and high-latitude Pacific Ocean. One example of the spatial extent of global ozone pollution is that the average concentration of ozone in remote areas of East Asia is already high enough to jeopardize agricultural and natural ecosystems there (14). It is easily seen, then, how the elevation of background levels of ozone by long-range transport can cause local or regional ozone, produced locally or regionally in amounts that would not otherwise have been critical, to exceed air-quality standards or critical levels (15, 16). This makes small

increments of ozone concentrations caused by contributions from other continents an issue of great concern (17).

The atmospheric lifetime of CO is also long enough (1 to 2 months on average) to allow intercontinental transport and hemispheric air pollution. Because a significant portion of CO pollution is from automobiles and biomass burning (13), its intercontinental transport is usually more easily captured by observation than that of ozone (18). Because the concentration of OH in remote areas is mainly controlled by CO, and the concentration of OH in the atmosphere determines the lifetimes of most atmospheric trace gases, including greenhouse gases such as CH_4 and HCFC (13), global pollution by CO is worrisome

Science, Vol. 298, no. 5601, 2106–2107, 13 December 2002

COUNTING THE COST OF LONDON'S KILLER SMOG
Richard Stone

LONDON—In December 1952, an acrid yellow smog settled on this city and killed thousands of people. The catastrophe, known as the "Big Smoke," was a turning point in efforts to clean up polluted air in cities across the Western world. It has taken half a century, though, for some of the fog to clear around the death toll from the roiling sulfurous clouds. New research suggests that the U.K. government might have underestimated the number of smog-related deaths by a factor of three.

Experts agree that the foul fog, which descended on London for a weekend in December 1952, killed roughly 4,000 people that month alone. But researchers are now sparring over the cause of death of another 8,000 Londoners in January and February 1953. Fresh analyses, debated at a conference here earlier this week to mark the 50th anniversary of the Big Smoke, suggest that these people succumbed to delayed effects of the smog or to lingering pollution. Other analyses insist that many of the "excess" deaths in early 1953 were caused by influenza, a view that the government has always supported. The debate reveals how much is unknown even today about the effects of smog, which continues to menace big cities, particularly in developing countries with weak air-pollution laws.

The Big Smoke. Victorian streets like this one were swept by an acrid cloud that killed thousands in London in late 1952.

Even in a city legendary for its "pea-soup" fogs, the Big Smoke is the stuff of legend. In early December 1952, an area of high pressure settled over London. Residents kept piling sulfur-rich coal into their stoves to keep warm in the near-freezing temperatures. In the still air, the smoke from these stoves and from coal-fired power plants in the city formed a smog laden with sulfur dioxide and soot. On Friday, 5 December, schools closed and transportation was disrupted. On Saturday night, a performance of the opera *La Traviata* had to be abandoned after smog obscured the stage. It wasn't until Tuesday, 9 December, that winds finally swept away the fouled air.

By then, it was clear that a disaster was unfolding, as scores upon scores of people succumbed to respiratory or heart ailments. In 1953, the Ministry of Health concluded that the deaths of 3,500 to 4,000 people—nearly three times the normal toll during such a period—could be attributed to the smog. But officials decreed that any deaths after 20 December had to be from other causes. During the first 3 months of 1953, there were 8,625 more deaths than expected. Officials put 5,655 down to flu and listed 2,970 as unexplained.

because of its effect on the oxidizing capacity of the atmosphere.

Another important aspect of global air pollution is the impact of aerosols on climate (7, 19). Aerosol lifetimes are approximately 1 to 2 weeks (19), which is significantly shorter than that of ozone. Therefore, aerosols have a more uneven distribution than ozone, both horizontally and vertically, and are more concentrated near their source regions over continents and in the boundary layer. The more uneven distribution of tropospheric aerosols causes highly heterogeneous radiative forcing (see page 92), which can lead to climate effects occurring regionally as well as globally (7, 19).

From the perspective of air quality, background concentrations of anthropogenic aerosols in remote areas are much lower than those considered dangerous by air-quality standards. This is because of their shorter lifetimes. Intercontinental transport of these aerosols is more episodic than for ozone.

Studies of transboundary air pollution led to the investigation of possible intercontinental transport (20) and hemispheric air pollution (21). Trans-Pacific transport of trace gases from Asia to North America has been reported most frequently (18, 22–26). Transport of Asian dust has been clearly identified in several events backed up by model analysis (27, 28). Although trans-Pacific transport of surface ozone has not been captured by observation, modeling studies have revealed that the Asian outflow enhances the concentration of surface ozone in the United States by a few ppbv (17).

Trans-Atlantic transport of O_3 and CO from North America to Europe during the period from 1990 to 1995 has been investigated (29) using data from Mace Head, on the west coast of Ireland, but relatively few episodes have been identified. Results from a chemical transport model and backward trajectories have shown that North American pollution contributes an average of approximately 5 ppbv to surface O_3 at Mace Head and about 2 to 4 ppbv over Europe in summer (up to 5 to 10 ppbv during some events) (30). The influence of North American pollution

KEY TERM

An **aerosol** is a dispersion of solid and liquid particles suspended in gas. Aerosols are produced by dozens of different processes that occur on land and water surfaces, and in the atmosphere itself. Aerosols occur in both the troposphere and the stratosphere, where they influence the amount of solar radiation reaching the surface of the Earth. The transport of pollutants via aerosols has significant ecosystem impacts.

on European air quality is seen most frequently in the free troposphere (31, 32).

Transport of European outflow across Eurasia to Asia has scarcely been studied. A study of backward trajectories has shown that a substantial amount of air from Europe arrives over East Asia in winter and early spring (33). Analyses of surface O_3 and CO data obtained at Mondy, a remote mountain site in eastern Siberia south of Lake Baikal, have shown that air masses transported from Europe have average concentrations of O_3 that are 2 to 3 ppbv higher, and of CO that are 6 to 14 ppbv higher, than those arriving from other regions (34). Surface measurements of trace gases over Siberia have been made using the trans-Siberian railroad between Moscow and Vladivostok (35, 36). Measurements of air pollutants over Eurasia made using commercial airlines have revealed high concentrations of O_3 in the upper troposphere (37, 38). Export of nitrogen oxides (NO_x), an important precursor of O_3 in the troposphere, from the polluted boundary layer and its production by lightning are the major sources over polluted continents and the clean ocean, respectively (39). Recent modeling studies by Wild et al. (40) have revealed unexpectedly that the impact of boundary layer ozone over Japan from

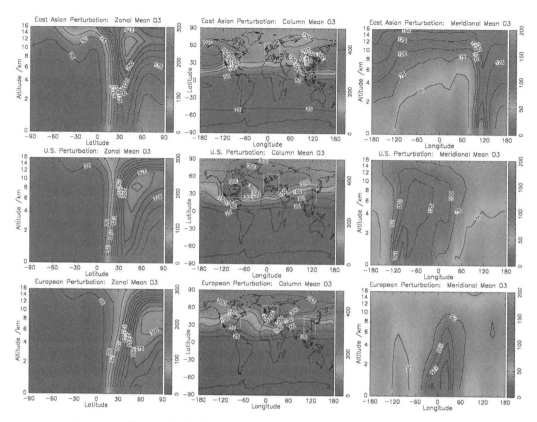

FIGURE 16.   Annual zonal (left column), column (center column), and meridional (right column) mean difference in O₃ mixing ratio (in ppbv) due to 10 percent increased emission of precursors over East Asia (top row), the United States (middle row), and Europe (bottom row) (20).

North America is comparable with that from Europe (0.2 to 2.5 ppbv). This greater North American impact is associated with a lifting of O_3 and its precursors into the upper troposphere over the Atlantic, and more efficient easterly transport of O_3 in the free troposphere. Wild and Akimoto (20) have studied the intercontinental transport and chemical transformation of O_3 between North America, Europe, and Asia using a global chemical transport model. Figure 16 shows the annual zonal, column, and meridional mean difference in O_3 mixing ratios (in ppbv) because of a 10 percent increase in emissions of three anthropogenic precursors of O_3—NO_x, CO, and volatile organic compounds—over East Asia, the United States, and Europe. The meridional mean values (Figure 16, right) highlight the elevated concentrations of O_3 above the polluted boundary layer and down-

wind of the region. Vertical transport processes move O_3 and its precursors emitted from East Asia close to the **tropopause** and effectively spread O_3

KEY TERM

The **tropopause** is the boundary layer between the troposphere and the stratosphere, varying in altitude from approximately 5 miles at the poles to near 11 miles at the equator. A team of scientists reported in the July 2003 issue of *Science* that human emissions are largely responsible for an increase in the height of the tropopause.

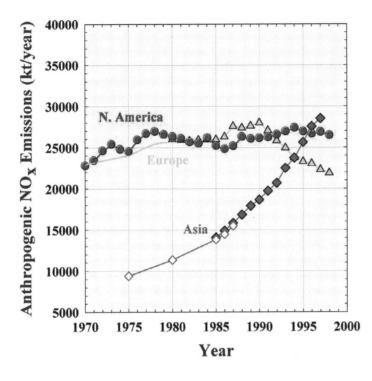

FIGURE 17. Changes in anthropogenic NO_x emissions over North America (United States and Canada) (43), Europe (including Russia and the near and middle East) (44), and Asia (East, Southeast, and South Asia) [solid squares, (46); open squares, (47)]. The extrapolated line for Europe in the 1970s is based on OECD data (45).

through the upper troposphere on a hemispheric scale, over North America and Europe as well. Thus, intercontinental transport of O_3 from East Asia occurs mostly in the middle and upper troposphere. In contrast, vertical transport of O_3 and its precursors is very weak in the case of European emissions, and downwind O_3 is confined to the boundary layer and middle troposphere. Thus, intercontinental transport of O_3 from Europe affects mainly near-surface O_3 concentrations in East Asia. European emissions produce the greatest enhancements over northern polar regions, whereas East Asian emissions occur sufficiently far south to affect the upper troposphere in the tropics and Southern Hemisphere as well (Figure 16, left). Emissions from the United States have an effect between that of East Asia and Europe for vertical, meridional, and zonal transport (Figure 16, middle row). Thus, the O_3 from the United States affects Europe in the boundary layer and middle and upper troposphere.

The hemispheric air pollution issue has recently been taken up by UN/ECE and U.S. EPA under the LATAP (Convention on Long-Range Transboundary Air Pollution), seeking future international agreement for mitigation (41). A first priority has been given to tackling pollution by ozone and aerosol, followed by mercury and POPs.

KEY TERMS

Air flows in different patterns according to altitude and orientation to the Earth. Swift and straight movement from a westerly point in an easterly direction is known as **zonal flow** (because the flow crosses many time/longitude zones in a short time). When flow buckles into large peaks and troughs, flow is said to be **meridional** (because the net movement is north/south along meridians of longitude over time).

Figure 17 shows the recent trend in NO_x emissions by continent in the Northern Hemisphere (42). Emissions from North America include those from the United States and Canada (43); European emissions include those from Russia and middle and near-East Asia (44, 45); and Asian emissions include those from East, Southeast, and South Asia (46, 47). Emissions from North America and Europe (including adjacent regions) have been nearly equal since the 1980s and have each remained near 25 to 28 Tg/year. After 1990, an apparently decreasing trend in NO_x emissions from Europe is thought to be the result of stringent emission controls in Western European countries. In contrast, Asian emissions, which contributed only a minor fraction of global emissions during the 1970s, have increased rapidly since then and surpassed emissions from North America and Europe in the mid-1990s.

This situation is expected to continue for at least the next couple of decades (48). In addition, future increases of emissions from Africa and South America, because of the economic growth there, would make global air quality more of an issue in the Southern Hemisphere, a region where only biomass burning has been considered important so far.

Finally, the importance of megacities as sources of regional and global pollution is worth noting. *Megacities* may be defined as metropolitan areas with over 10 million inhabitants, although there is no precise accepted threshold, and population estimates are not necessarily based on the same areas of reference. In 2001, there were 17 megacities according to United Nations statistics (49). With rapid growth of the world's population, particularly in developing countries, and continuing industrialization and migration toward urban centers, megacities are becoming more important sources of air pollution from associated mobile and stationary sources. Air quality in megacities is thus of great concern, as illustrated by a study in Mexico City (50). Although the health effects of air pollution on the inhabitants of megacities are a serious social problem, its regional and global environmental consequences are also of great concern.

Because of all these considerations, local, regional, and global air-quality issues, as well as regional and global environmental effects—including climate change—should be viewed in an integrated manner.

References

1. H. G. Reichle Jr. *et al.*, *J. Geophys. Res.* 91, 10865 (1986).

2. P. J. Crutzen *et al.*, *Nature* 282, 253 (1979).

3. J. P. Fishman, H. Reichle, *J. Geophys. Res.* 91, 14451 (1986).

4. P. Borrell *et al.*, *Atmos. Environ.* 37, 2567 (2003).

5. J.-F. Lamarque *et al.*, *Geophys. Res. Lett.* 30, 1688 (2003).

6. D. P. Edwards *et al.*, *J. Geophys. Res.* 108, 1029/2002JD002927 (2003).

7. V. Ramanathan *et al.*, *Science* 294, 2119 (2001).

8. Y. J. Kaufman, D. Tanre, O. Boucher, *Nature* 419, 215 (2002).

9. T. Nakajima *et al.*, *Geophys. Res. Lett.* 28, 1171 (2001).

10. A. Higurashi, T. Nakajima, *Geophys. Res. Lett.* 29, 10.1029/2002GL015357 (2002).

11. J. T. Houghton *et al.*, Eds., *Climate Change 2001: The Scientific Bases* (Cambridge Univ. Press, Cambridge, 2001), pp. 260–263.

12. G. Brasseur, J. J. Orlando, G. S. Tyndall, Eds., *Atmospheric Chemistry and Global Change* (Oxford Univ. Press, Oxford, 1999).

13. J. Lelieveld, F. Dentener, *J. Geophys. Res.* 105, 3531 (2000).

14. P. Pochanart *et al.*, *Atmos. Environ.* 36, 4235 (2002).

15. World Health Organization (WHO), *Update and Revision of the WHO Air Quality Guidelines for Europe. Ecotoxic Effects, Ozone Effects on Vegetation* (European Center for Environment and Health, Bilthoven, Netherlands, 2000).

16. R. C. Musselman, P. M. McCool, A. L. Lefohn, *J. Air, Waste Manag. Assoc.* 44, 1383 (1994).

17. D. L. Jacob, J. A. Logan, P. P. Murti, *Geophys. Res. Lett.* 26, 2175 (1999).

18. D. Jaffe *et al.*, *J. Geophys. Res.* 106, 7449 (2001).

19. J. Haywood, O. Boucher, *Rev. Geophys.* 38, 513 (2000).

20. O. Wild, H. Akimoto, *J. Geophys. Res.* 106, 27729 (2001).

21. T. Holloway, A. M. Fiore, M. G. Hastings, *Environ. Sci. Technol.*, 37, 4535 (2003).

22. D. D. Parrish *et al.*, *J. Geophys. Res.* 97, 15883 (1992).

23. D. Jaffe *et al.*, *Geophys. Res. Lett.* 26, 711 (1999).

24. D. Jaffe *et al.*, *Atmos. Environ.* 37, 391 (2003).

25. T. K. Bernstein, S. Karlsdottir, D. A. Jaffe, *Geophys. Res. Lett.* 26, 2171 (1999).

26. K. Wilkening, L. Barrie, M. Engle, *Science* 290, 65 (2000).

27. R. B. Husar *et al.*, *J. Geophys. Res.* 106, 18317 (2001).

28. I. Uno *et al.*, *J. Geophys. Res.* 106, 18331 (2001).

29. R. G. Derwent *et al.*, *Atmos. Environ.* 32, 145 (1998).

30. Q. Li *et al.*, *J. Geophys. Res.* 107, 10.1029/2001JD001422 (2002).

31. A. Stohl, T. Trickl, *J. Geophys. Res.* 104, 30445 (1999).

32. A. Stohl *et al.*, *J. Geophys. Res.* 106, 1029/2001JD001396 (2002).

33. R. E. Newell, M. J. Evans, *Geophys. Res. Lett.* 27, 2509 (2000).

34. P. Pochanart *et al.*, *J. Geophys. Res.* 108, 10.1029/2001JD001412 (2003).

35. P. Bergamaschi *et al.*, *J. Geophys. Res.* 103, 8227 (1998).

36. T. Roeckman *et al.*, *Chemosphere Global Change Sci.* 1, 219 (1999).

37. D. Brunner *et al.*, *J. Geophys. Res.* 106, 27673 (2001).

38. A. Stohl *et al.*, *J. Geophys. Res.* 106, 27757 (2001).

39. D. P. Jeker *et al.*, *J. Geophys. Res.* 105, 3679 (2000).

40. O. Wild, P. Pochanart, H. Akimoto, *J. Geophys. Res.* 109, 10.1029/2003JD004501 (2004).

41. United Nations Economic and Social Council and U.S. Environmental Protection Agency, First Task Force on Hemispherical Transport of Air Pollution, Brussels, 1 to 3 June 2005.

42. M. Naja, H. Akimoto, J. Staehelin, *J. Geophys. Res.* 108, 10.1029/2002JD002477 (2003).

43. U.S. Environmental Protection Agency (EPA), *National Air Pollutant Emission Trends, 1900–1998* (Report EPA-454/R-00-002, EPA, Research Triangle Park, NC, 1999).

44. Co-operative Programme for Monitoring and Evaluation of the Long-Range Transmission of Air Pollutants in Europe (EMEP), "Emission data reported to UNECP/ EMEP: Evaluation of the spatial distribution of emissions, MSC_W" (status report 2001 by V. Vestreng, Norwegian Meteorological Institute, Oslo, Norway, 2001).

45. Organisation for Economic Co-operation and Development (OECD), *OECD Environmental Data Compendium 1993* (OECD, Paris, France, 1993).

46. N. Kato, H. Akimoto, *Atmos. Environ.* 26, 2997 (1992).

47. D. Streets *et al.*, *Water Air Soil Pollution* 130, 187 (2001).

48. Z. Klimont *et al.*, *Water Air Soil Pollution* 130, 193 (2001).

49. United Nations, *World Urbanization Prospects, the 2001 Revision* (United Nations, Population Division, Department of Economic and Social Affairs, New York, 2002).

50. L. T. Molina, M. J. Molina, *Air Quality in the Mexico Megacity* (Kluwer Academic, Dortrecht, Netherlands, 2002).

Modern Global Climate Change

THOMAS R. KARL and KEVIN E. TRENBERTH

odern climate change is dominated by human influences,
which are now large enough to
exceed the bounds of natural variability. The
main source of global climate change is human-
induced changes in atmospheric composition.
These perturbations primarily result from emissions associated with energy use, but on local and
regional scales, urbanization and land use
changes are also important. Although there has
been progress in monitoring and understanding
climate change, there remain many scientific,
technical, and institutional impediments to precisely planning for, adapting to, and mitigating
the effects of climate change. There is still considerable uncertainty about the rates of change
that can be expected, but it is clear that these

changes will be increasingly manifested in
important and tangible ways, such as changes in
extremes of temperature and precipitation,
decreases in seasonal and perennial snow and ice
extent, and sea level rise. Anthropogenic climate
change is now likely to continue for many centuries. We are venturing into the unknown with
climate, and its associated effects could be quite
disruptive.

The atmosphere is a global commons that
responds to many types of emissions into it, as
well as to changes in the surface beneath it. As
human balloon flights around the world illustrate, the air over a specific location is typically
halfway around the world a week later, making
climate change a truly global issue.

Planet Earth is habitable because of its location relative to the sun and because of the natural greenhouse effect of its atmosphere.
Various atmospheric gases contribute to the
greenhouse effect, the impact of which in clear

This article first appeared in *Science* (5
December 2003: Vol. 302, no. 5651). It has been
revised and updated for this edition.

KEY TERM

A **proxy** climate indicator is a local record that is interpreted, using physical and biophysical principles, to represent some combination of climate-related variations back in time. Climate-related data derived in this way are referred to as **proxy data**. Examples of proxies are tree ring records, characteristics of corals, and various data derived from ice cores.

the use of climate models for assessing the past and making projections into the future, and the need for better observational and information systems.

The main way in which humans alter global climate is by interferring with the natural flows of energy through changes in atmospheric composition, not by actually generating heat in energy usage. On a global scale, even a 1 percent change in the energy flows, which is the order of the estimated change to date (2), dominates all other direct influences humans have on climate. For example, an energy output of just 1 PW is equivalent to that of a million power stations of 1000-MW capacity, among the largest in the world. Total human energy use is about a factor of 9000 less than the natural flow (3).

Global changes in atmospheric composition occur from anthropogenic emissions of greenhouse gases, such as carbon dioxide that results from burning fossil fuels and methane and nitrous oxide from multiple human activities. Because these gases have long (decades to centuries) atmospheric lifetimes, the result is an accumulation in the atmosphere and a buildup in concentrations that are clearly shown both by instrumental observations of air samples since 1958 and in bubbles of air trapped in ice cores before then. Moreover, these gases are well distributed in the atmosphere across the globe, simplifying a global monitoring strategy. Carbon dioxide has increased 31 percent since preindustrial times, from 280 parts per million by volume (ppmv) to more than 370 ppmv today, and half of the increase has been since 1965 (4) (figure 18). The greenhouse gases trap outgoing

skies is ~60 percent from water vapor, ~25 percent from carbon dioxide, ~8 percent from ozone, and the rest from trace gases including methane and nitrous oxide (1). Clouds also have a greenhouse effect. On average, the energy from the sun received at the top of the Earth's atmosphere amounts to 175 petawatts (PW) (or 175 quadrillion watts), of which ~31 percent is reflected by clouds and from the surface. The rest (120 PW) is absorbed by the atmosphere, land, or ocean and ultimately emitted back to space as infrared radiation (1). Over the past century, infrequent volcanic eruptions of gases and debris into the atmosphere have significantly perturbed these energy flows; however, the resulting cooling has lasted for only a few years (2). Inferred changes in total **solar irradiance** appear to have increased global mean temperatures by perhaps as much as 0.2°C in the first half of the 20th century, but measured changes in the past 25 years are small (2). Over the past 50 years, human influences have been the dominant detectable influence on climate change (2). The following briefly describes the human influences on climate, the resulting temperature and precipitation changes, the time scale of responses, some important processes involved,

KEY TERM

Solar irradiance is the total solar radiation measured at mean Sun–Earth distance.

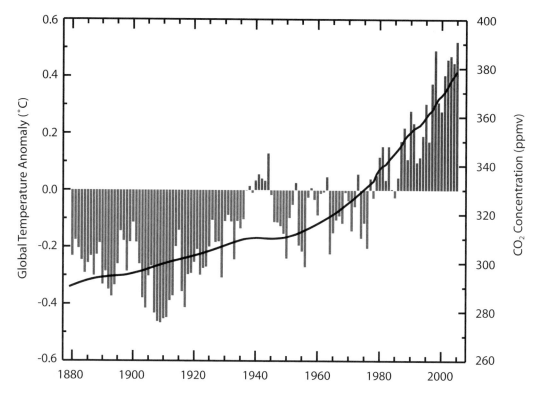

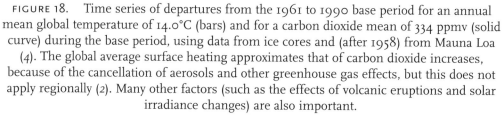

FIGURE 18. Time series of departures from the 1961 to 1990 base period for an annual mean global temperature of 14.0°C (bars) and for a carbon dioxide mean of 334 ppmv (solid curve) during the base period, using data from ice cores and (after 1958) from Mauna Loa (4). The global average surface heating approximates that of carbon dioxide increases, because of the cancellation of aerosols and other greenhouse gas effects, but this does not apply regionally (2). Many other factors (such as the effects of volcanic eruptions and solar irradiance changes) are also important.

radiation from the Earth to space, creating a warming of the planet.

Emissions into the atmosphere from fuel burning further result in gases that are oxidized to become highly reflective micron-sized aerosols, such as sulfate, and strongly absorbing aerosols, such as black carbon or soot. Aerosols are rapidly (usually 2 weeks or less) removed from the atmosphere through the natural hydrological cycle and dry deposition as they travel away from their source. Nonetheless, atmospheric concentrations can substantially exceed background conditions in large areas around and downwind of the emission sources. Depending on their reflectivity and absorption properties,

geometry and size distribution, lifetimes in the atmosphere, and interactions with clouds and moisture, these particulates can lead to either net cooling, as for sulfate aerosols, or net heating, as for black carbon. Importantly, sulfate aerosols affect climate directly by reflecting solar radiation and indirectly by changing the reflective properties of clouds and their lifetimes. Understanding their precise impact has been hampered by our inability to measure these aerosols directly, as well as by their spatial inhomogeneity and rapid changes in time. Large-scale measurements of aerosol patterns have been inferred through emission data, special field experiments, and indirect measurements such as sun photometers (5).

SCIENCE IN THE NEWS

Science, Vol. 291, no. 5509, 1690–1691, 2 March 2001

THE MELTING SNOWS OF KILIMANJARO
Robert Irion

SAN FRANCISCO—"As wide as all the world, great, high, and unbelievably white in the sun, was the square top of Kilimanjaro." Those evocative words by Ernest Hemingway describe a scene that could vanish within 20 years, according to new field research. More than 80 percent of the ice on Africa's highest peak has melted since the early 20th century, joining other glaciers that are ebbing from the world's tropical mountains at an accelerating rate.

The dramatic findings, splashed on front pages and the evening news, may spur policymakers and the public far more than abstract warnings of climatic trends, says Will Steffen, director of the International Geosphere-Biosphere Program in Stockholm, Sweden. "This is exceptionally important work," Steffen says. "Tropical glaciers are a bellwether of human influence on the Earth system." The past decade's warm years, it seems, have sounded that bell with unexpected force.

Ice in the tropics sits at the knife edge of climate change. Slight temperature increases push the snowline to ever-higher altitudes, saturating fields of ice with water. Glaciers, which normally drain slowly from an ice cap and maintain a steady size, begin to melt and retreat. Researchers have observed ice waning on peaks in Kenya, Venezuela, New Guinea, Ecuador, and elsewhere. The famous ice fields on Kilimanjaro and in Peru appear especially frail.

Ebbing Ice. There has been an 82 percent decline in glacier size since 1912 on Mount Kilimanjaro's Uhuru Peak.
CREDIT: OGEN PERRY/ISTOCKPHOTO.COM

Aerial mapping of Kilimanjaro's summit in February 2000 revealed a 33 percent loss of ice since the last map in 1989 and an 82 percent decline since 1912, says geologist Lonnie Thompson of Ohio State University's Byrd Polar Research Center in Columbus. Just 2 weeks ago, Thompson's colleagues measured the levels at survey poles that they inserted into the ice pack last year. More than a meter of ice had melted in 12 months, out of a total thickness of 20 to 50 meters. "It won't take many more years like that to completely melt the ice fields," Thompson says.

Moreover, Thompson's group has documented runaway melting at Quelccaya, a massive ice cap in the Andes of Peru. Surveys reveal that Qori Kalis, Quelccaya's main drainage glacier, has retreated 155 meters per year since 1998. That's 32 times faster than the rate between 1963 and 1978. The area of the ice cap itself has shrunk from 56 square kilometers in 1976 to 44 square kilometers today. The hastening pace suggests that it too may dribble away within 20 years, says Thompson.

Human activities also have a large-scale impact on the land surface. Changes in land use through urbanization and agricultural practices, although not global, are often most pronounced where people live, work, and grow food, and are part of the human impact on climate (6, 7). Large-scale deforestation and desertification in Amazonia and the Sahel, respectively, are two instances where evidence suggests there is likely to be human influence on regional climate (8–10). In general, city climates differ from those in surrounding rural green areas, because of the "concrete jungle" and its effects on heat retention, runoff, and pollution, resulting in urban heat islands.

There is no doubt that the composition of the atmosphere is changing because of human activities, and today greenhouse gases are the largest human influence on global climate (2). Recent greenhouse gas emission trends in the United States are upward (11), as are global emissions trends, with increases between 0.5 and 1 percent per year over the past few decades (12). Concentrations of both reflective and nonreflective aerosols are also estimated to be increasing (2). Because radiative forcing from greenhouse gases dominates over the net cooling forcings from aerosols (2), the popular term for the human influence on global climate is "global warming," although it really means global heating, of which the observed global temperature increase is only one consequence (13) (Figure 18). Already it is estimated that the Earth's climate as reflected in the global mean temperature has exceeded the bounds of natural variability (2). This has been the case since about 1980.

Surface moisture, if available (as it always is over the oceans), effectively acts as the "air conditioner" of the surface, as heat used for evaporation moistens the air rather than warming it. Therefore, another consequence of global heating of the lower troposphere is accelerated land-surface drying and more atmospheric water vapor (the dominant greenhouse gas) (14). Accelerated drying increases the incidence

and severity of droughts (15), whereas additional atmospheric water vapor increases the risk of heavy precipitation events (16). Basic theory (17), climate model simulations (2), and empirical evidence (Figure 19) all confirm that warmer climates, owing to increased water vapor, lead to more intense precipitation events even when the total precipitation remains constant, and with prospects for even stronger events when precipitation amounts increase (18–21).

There is considerable uncertainty as to exactly how anthropogenic global heating will affect the climate system, how long it will last, and how large the effects will be. Climate has varied naturally in the past, but today's circumstances are unique because of human influences on atmospheric

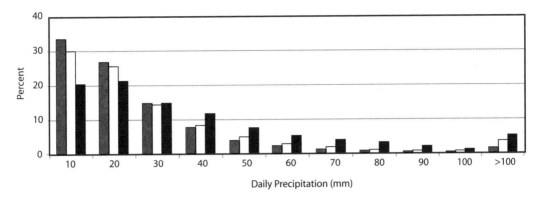

FIGURE 19.   Climatology of the intensity of daily precipitation as a percentage of total amount in 10 mm/day categories for different temperature regimes, based on 51, 37, and 12 worldwide stations, respectively: blue bars, −3°C to 19°C; pink bars, 19°C to 29°C; dark red bars, 29°C to 35°C. By selection, all stations have the same seasonal mean precipitation amount of 230 (5 mm). As temperatures and the associated water-holding capacity of the atmosphere (15) increase, more precipitation falls in heavy (more than 40 mm/day) to extreme (more than 100 mm/day) daily amounts.

composition. As we progress into the future, the magnitude of the present anthropogenic change will become overwhelmingly large compared with that of natural changes. In the absence of climate mitigation policies, the 90 percent probability interval for warming from 1990 to 2100 is 1.7° to 4.9°C (22). About half of this range is due to uncertainty in future emissions and about half is due to uncertainties in climate models (2, 22), especially in their sensitivity to forcings that are complicated by feedbacks, discussed below, and in their rate of heat uptake by the oceans (23). Even with these uncertainties, the likely outcome is more frequent heat waves, droughts, extreme precipitation events, and related impacts (such as wildfires, heat stress, vegetation changes, and sea level rise) that will be regionally dependent.

The rate of human-induced climate change is projected to be much faster than most natural processes, certainly those prevailing over the past 10,000 years (2). Thresholds likely exist that, if crossed, could abruptly and perhaps almost irreversibly switch the climate to a different regime. Such rapid change is evident in past climates during a slow change in the Earth's orbit

and tilt, such as the Younger Dryas cold event from ~11,500 to ~12,700 years ago (2), perhaps caused by freshwater discharges from melting ice sheets into the North Atlantic Ocean and a change in the ocean thermohaline circulation (24, 25). The great ice sheets of Greenland and Antarctica may not be stable, because the extent

KEY TERM

Thermohaline circulation is a large-scale density-driven circulation in the ocean, caused by differences in temperature and salinity. In the North Atlantic it consists of warm surface water flowing northward and cold water flowing southward, resulting in a net poleward transport of heat. The surface water sinks to depth in highly restricted sinking regions located in high latitudes.

to which cold-season heavier snowfall partially offsets increased melting as the climate warms remains uncertain. A combination of ocean temperature increases and ice sheet melting could systematically inundate the world's coasts by raising sea level for centuries.

Given what has happened to the climate to date and is projected to happen in the future (2), substantial further climate change is guaranteed. The rate of change can be slowed, but it is unlikely to be stopped in the 21st century (26). Because concentrations of long-lived greenhouse gases are dominated by accumulated past emissions, it takes many decades for any change in emissions to have much effect. This means the atmosphere still has unrealized warming (estimated to be at least another 0.5°C) and that sea level may continue rise for centuries after anthropogenic greenhouse gas emissions have abated and greenhouse gas concentrations in the atmosphere have stabilized.

Our understanding of the climate system is complicated by feedbacks that either amplify or damp perturbations. The most important of these involve water in various phases. As temperatures increase, the water-holding capacity of the atmosphere increases along with water vapor amounts, producing water vapor feedback. As water vapor is a strong greenhouse gas, this diminishes the loss of energy through infrared radiation to space. Currently, water vapor feedback is estimated to contribute a radiative effect from one to two times the size of the direct effect of increases in anthropogenic greenhouse gases (27, 28). Precipitation-runoff feedbacks occur because more intense rains run off at the expense of soil moisture, and warming promotes rain rather than snow. These changes in turn alter the partitioning of solar radiation into **sensible** versus **latent heating** (16). Heat storage feedbacks include the rate at which the oceans take up heat and the currents redistribute and release it back into the atmosphere at variable later times and different locations.

Cloud feedback occurs because clouds both reflect solar radiation, causing cooling, and trap

KEY TERMS

Sensible heating is the heat absorbed or released by a molecule associated with a change in temperature, but without any gain or loss of heat that may accompany a change in state.

Latent heating is the heat absorbed or released by a molecule in a reversible phase change, such as from liquid water (rain) to water vapor.

outgoing long-wave radiation, causing warming. Depending on the height, location, and type of clouds with their related optical properties, changes in cloud amount can cause either warming or cooling. Future changes in clouds are the single biggest source of uncertainty in climate predictions. Possible changes in clouds contribute to an uncertainty in the sensitivity of models to changes in greenhouse gases. Cloud feedback effects range from a small negative feedback, thereby slightly reducing the direct radiative effects of increases in greenhouse gases, to a doubling of the direct radiative effect of increases in greenhouse gases (28). Clouds and precipitation processes cannot be resolved in climate models and have to be parametrically represented (parameterized) in terms of variables that are resolved. This will continue for some time into the future, even with projected increases in computational capability (29).

Ice-albedo feedback occurs as increased warming diminishes snow and ice cover, making the planet darker and more receptive to absorbing incoming solar radiation, causing warming, which further melts snow and ice. This effect is greatest at high latitudes. Decreased snow cover extent has significantly contributed to the earlier onset of spring in the past few decades over

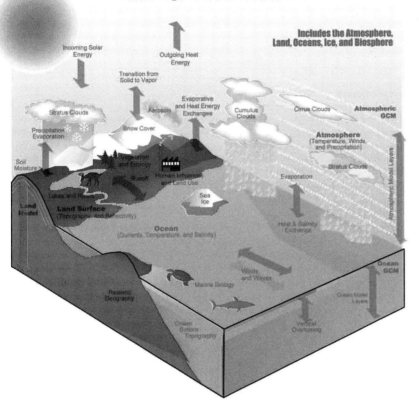

Modeling the Climate System

Includes the Atmosphere, Land, Oceans, Ice, and Biosphere

FIGURE 20.
Components of the climate system and the interactions among them, including the human component. All these components have to be modeled as a coupled system that includes the oceans, atmosphere, land, cryosphere, and biosphere. GCM, General Circulation Model.

Northern-Hemisphere high latitudes (30). Ice-albedo feedback is affected by changes in clouds, thus complicating the net feedback effect.

The primary tools for predicting future climate are global climate models, which are fully coupled, mathematical, computer-based models of the physics, chemistry, and biology of the atmosphere, land surface, oceans, and cryosphere and their interactions with each other and with the sun and other influences (such as volcanic eruptions). Outstanding issues in modeling include specifying forcings of the climate system; properly dealing with complex feedback processes (Figure 20) that affect carbon, energy, and water sources, sinks and transports; and improving simulations of regional weather, especially extreme events. Today's inadequate or incomplete measurements of various forcings, with the exception of well-mixed greenhouse gases, add uncertainty when trying to simulate past and present climate. Confidence in our ability to predict future climate depends on our ability to use climate models to attribute past and present climate change to specific forcings. Through clever use of paleoclimate data, our ability to reconstruct past forcings should improve, but it is unlikely to provide the regional detail necessary that comes from long-term direct measurements. An example of forcing uncertainty comes from recent satellite observations and data analyses of 20th-century surface, upper air, and ocean temperatures, which indicate that estimates of the indirect effects of sulfate aerosols on clouds may be lower than earlier thought, perhaps by as much as a factor of two (*31–33*). Human behavior, technological change, and the rate of population growth also affect future emissions and our ability to predict these must be factored into any long-term climate projection.

Regional predictions are needed for improving assessments of vulnerability to and impacts of change. The coupled atmosphere-ocean system has a preferred mode of behavior known as El Niño, and similarly the atmosphere is known to have preferred patterns of behavior, such as the North Atlantic Oscillation (NAO). So how will El Niño and the NAO change as the climate changes? There is evidence that the NAO, which affects the severity of winter temperatures and precipitation in Europe and eastern North America, and El Niño, which has large regional effects around the world, are behaving in unusual ways that appear to be linked to global heating (*2, 34–36*). Hence, it is necessary to be able to predict the statistics of the NAO and El Niño to make reliable regional climate projections.

Ensembles of model predictions have to be run to generate probabilities and address the chaotic aspects of weather and climate. Probabilities derived from ensembles of models can be addressed in principle with adequate computing power, a challenge in itself. However, improving models to a point where they are more reliable and have sufficient resolution to be properly able to represent known important processes also requires the right observations, understanding, and insights (brain power). Global climate models will need to better integrate the biological, chemical, and physical components of the Earth system (Figure 20). Even more challenging is the seamless flow of data and information among observing systems, Earth system models, socio-economic models, and models that address managed and unmanaged ecosystems. Progress here depends on overcoming not only scientific and technical issues but also major institutional and international obstacles related to the free flow of climate-related data and information.

In large part, reduction in uncertainty about future climate change will be driven by studies of climate change assessment and attribution. Along with climate model simulations of past climates, this requires comprehensive and long-term climate-related data sets and observing systems that deliver data free of time-dependent biases. These observations would ensure that model simulations are evaluated on the basis of actual changes in the climate system and not on artifacts of changes in observing system technology or analysis methods (*37*). Ongoing controversies, such as the effects of changes in observing systems related to the rate of surface versus tropospheric warming (*38, 39*) highlights this issue. Global monitoring through space-based and surface-based systems is an international matter, much like global climate change. There are encouraging signs, such as the adoption in 1999 of a set of climate monitoring principles (*40*), but these principles are impotent without implementation. International implementation of these principles is spotty at best (*41*).

KEY TERM

El Niño, in its original sense, was a warm water current that periodically flows along the coast of Ecuador and Peru, disrupting the local fishery. It has since become identified with a basinwide warming of the tropical Pacific east of the dateline. This oceanic event is associated with a fluctuation of a global scale tropical and subtropical surface pressure pattern, called the Southern Oscillation. This coupled atmosphere-ocean phenomenon is collectively known as El Niño-Southern Oscillation, or ENSO. During an El Niño event the prevailing trade winds weaken, reducing upwelling and altering ocean currents such that sea surface temperatures warm, which further reduce the trade winds. This event has great impact on the wind, sea surface temperature, and precipitation patterns in the tropical Pacific. It has climatic effects throughout the Pacific region and in many other parts of the world. The cold phase of ENSO is called La Niña.

We are entering the unknown with our climate. We need a global climate observing system, but only parts of it exist. We must not only take the vital signs of the planet but also assess why they are fluctuating and changing. Consequently, the system must embrace comprehensive analysis and assessment as integral components on an ongoing basis, as well as innovative research to better interpret results and improve our diagnostic capabilities. Projections into the future are part of such activity, and all aspects of an Earth information system feed into planning for the future, whether by planned adaptation or mitigation. Climate change is truly a global issue, one that may prove to be humanity's greatest challenge. It is very unlikely to be adequately addressed without greatly improved international cooperation and action.

References and Notes

1. J. T. Kiehl, K. E. Trenberth, *Bull. Am. Meteorol. Soc.* 78, 197 (1997).

2. J. T. Houghton *et al.*, Eds., *Climate Change 2001: The Scientific Basis* (Cambridge Univ. Press, Cambridge, 2001) (available at www.ipcc.ch/).

3. R. J. Cicerone, *Proc. Natl. Acad. Sci. U.S.A.* 100, 10304 (2000).

4. Atmospheric CO_2 concentrations from air samples and from ice cores are available at http://cdiac.esd.ornl.gov/trends/co2/sio-mlo.htm and http://cdiac.esd.ornl.gov/trends/co2/siple.htm, respectively.

5. M. Sato *et al.*, *Proc. Natl. Acad. Sci. U.S.A.* 100, 6319 (2003).

6. A. J. Dolman, A. Verhagen, C. A. Rovers, Eds. *Global Environmental Change and Land Use* (Kluwer, Dordrecht, Netherlands, 2003).

7. G. B. Bonan, *Ecol. Appl.* 9, 1305 (1999).

8. J. G. Charney, *Q. J. R. Meteorol. Soc.* 101, 193 (1975).

9. C. Nobre *et al.*, in *Vegetation, Water, Humans, and the Climate*, P. Kabot *et al.*, Eds. (Springer Verlag, Heidelberg, Germany, 2004).

10. A. N. Hahmann, R. E. Dickinson, *J. Clim.* 10, 1944 (1997).

11. U.S. Department of State, *U.S. Climate Action Report 2002* (Washington, DC, 2002) (available at http://yosemite.epa.gov/oar/globalwarming.nsf/content/ResourceCenterPublicationsUSClimateActionReport.html).

12. G. Marland, T. A. Boden, R. J. Andres, at the Web site *Trends: A Compendium of Data on Global Change* (CO_2 Information Analysis Center, Oak Ridge National Laboratory, Oak Ridge, TN, 2002; available at http://cdiac.esd.ornl.gov/trends/emis/em-cont.htm).

13. Global temperatures are available at www.ncdc .noaa.gov/oa/climate/research/anomalies/anomalies .html.

14. K. E. Trenberth, J. Fasullo, L. Smith, *Clim. Dyn.* 24, 741 (2005).

15. A. Dai, K. E. Trenberth, T. Qian, *J. Hydrometeor.* 5, 1117 (2004).

16. K. E. Trenberth, A. Dai, R. M. Rasmussen, D. B. Parsons, *Bull. Am. Meteorol. Soc.* 84, 1205 (2003) (available at www.cgd.ucar.edu/cas/trenberth.pdf/ rainChBamsR.pdf).

17. The Clausius Clapeyron equation governs the waterholding capacity of the atmosphere, which increases by ~7 percent per degree Celsius increase in temperature (*16*).

18. R. W. Katz, *Adv. Water Res.* 23, 133 (1999).

19. P. Ya. Groisman, *Clim. Change.* 42, 243 (1999).

20. T. R. Karl, R. W. Knight, *Bull. Am. Meteorol. Soc.* 78, 1107 (1998).

21. P. Ya. Groisman, R.W. Knight, D. R. Easterling, T. R. Karl, G. C. Hagerl, *J. Clim.* 18, 9 (2005).

22. T. Wigley, S. Raper, *Science* 293, 451 (2001).

23. S. J. Levitus *et al.*, *Science* 287, 2225 (2001).

24. W. S. Broecker, *Science* 278, 1582 (1997).

25. T. F. Stocker, O. Marchal, *Proc. Natl. Acad. Sci. U.S.A.* 97, 1362 (2000).

26. M. Hoffert *et al.*, *Science* 298, 981 (2002).

27. U.S. National Research Council, *Climate Change Science: An Analysis of Some Key Questions* (National Academy, Washington, DC, 2001).

28. R. Colman, *Clim. Dyn.* 20, 865 (2003).

29. T. R. Karl, K. E. Trenberth, *Sci. Am.* 281, 100 (December 1999).

30. P. Ya. Groisman, T. R. Karl, R. W. Knight, G. L. Stenchikov, *Science* 263, 198 (1994).

31. C. E. Forest, P. H. Stone, A. Sokolov, M. R. Allen, M. D. Webster, *Science* 295, 113 (2002).

32. J. Coakley Jr., C. D. Walsh, *J. Atmos. Sci.* 59, 668 (2002).

33. J. Coakley Jr., personal communication.

34. M. A. Saunders, *Geophys. Res. Lett.* 30, 1378 (2003).

35. M. P. Hoerling, J. W. Hurrell, T. Xu, *Science* 292, 90 (2001).

36. K. E. Trenberth, T. J. Hoar, *Geophys. Res. Lett.* 24, 3057 (1997).

37. K. E. Trenberth, T. R. Karl, T. W. Spence, *Bull. Amer. Meteor. Soc.* 83, 1558 (2002).

38. The Climate Change Science Program plan is available at www.climatescience.gov.

39. B. Santer *et al.*, *Science*, 300, 1280 (2003).

40. The climate principles were adopted by the Subsidiary Body on Science, Technology and Assessment of the United Nations Framework Convention on Climate Change (UNFCCC).

41. Global Climate Observing System (GCOS), *The Second Report on the Adequacy of the Global Observing Systems for Climate in Support of the UNFCCC.* (GCOS-82, WMO/TD 1143, World Meteorological Organisation, Geneva, 2003) (available from www.wmo.ch/web/gcos/ gcoshome.html).

42. We thank A. Leetmaa, J. Hurrell, J. Mahlman, and R. Cicerone for helpful comments, and J. Enloe for providing the calculations for Figure 19. This chapter reflects the views of the authors and does not reflect government policy. The National Climatic Data Center is part of NOAA's Satellite and Information Services. The National Center for Atmospheric Research is sponsored by the NSF.

Web Resources

www.sciencemag.org/cgi/content/full/302/5651/1719/ DC1

The Commons

Managing Our Common Inheritance

DONALD KENNEDY

Thirty-eight years ago, *Science* published a remarkable essay by Garret Hardin entitled "The Tragedy of the Commons." Those who knew Hardin at the time and admired his paper had no idea whatsoever of the influence it and its author would have on how we think about population and the environment. That influence has spawned several successor strands. One, evident almost immediately, was an enhanced concern about the impact of population growth on resource utilization. The second was a delayed argument, emerging in the 1980s, about how to consider population growth in policy terms, and a notable outcome of that was a dramatically growing movement toward population control and reproductive health—an interest than continues, with notable successes, today. The third, much later, is a recent social science literature revising Hardin's "hard choices" by showing that groups often evolve fair social arrangements that limit exploitation and conserve shared resources.

Before turning to that and other sequels to the original "Tragedy," the reader should examine Hardin's *Science* essay carefully. What "hard choices" are proposed, and why? Hardin begins with a discouraging point about the unavailability of purely "technical" solutions, arguing by analogy with the problem of nuclear détente during the Cold War. That is also true, his argument runs, of the population problem; Hardin draws on Malthus and others to show that the optimal population is less than the maximum population, and that it is unlikely that the optimum can be gotten with complete reproductive freedom. He goes on to present a choice between two freedoms: the freedom to breed, and the freedom to exploit the commons. Limitations, he argues, must be placed on one or the other: Get rid of the commons by stern governmental regulation of access, or privatization; or limit the number of children people may have.

Well, where have we come in 38 years? The population/resource collision has only grown more important since Hardin's *Science* essay. Earth's population then was about 3.5 billion; it has since grown by a factor of nearly two, to 6.3 billion. That growth, amplified by global increases in affluence and the power of technology, has brought escalating pressures on commons resources, now often called "common pool" resources—such as air, fresh water, and ocean fisheries. These have in common that they are accessible to many potential harvesters who can extract marginal personal benefits at a cost that is low because all other harvesters share it. Decades of depletion of these resources, whose status was explored in the preceding two sections of this book, have led to new concerns and new terms: *sustainability* and *sustainability science*. The loss of resource value compels us to undertake more careful analyses: first, of what values we actually take from nature's resources, and second, of how science can contribute to maintaining such resources sustainably. The essays in this section will explore an extensive literature that has arisen in reaction to "Tragedy," and it offers some encouragement about sustainability. The cases examined by these social scientists involve local or even regional groups of harvesters who organize cooperative rules that ensure that the commons is preserved.

The notion of sustainability and preservation of the commons depends on the way in which we value the environment—a global concept, obviously, that includes "commons" of many different kinds. We obtain value in various ways: We may use the environment for timber or for hunting, we may enjoy it for various nonuse values such as bird watching, and we may extract pleasure from merely knowing that it's there. In "Man and Nature," perhaps the first environmental classic, George Perkins Marsh provided a meticulous 19th-century account of what had happened to the world's woods, waters, and fields. In Marsh one finds a kind of outrage over environmental damage, but there is little of the sense of wonder about nature that one finds in modern writers such as Wallace Stegner. Marsh is all about use values, Stegner about nonuse. A modern convergence defines sustainability as requiring that the average welfare of the successor generation, with respect to the total of all these values, be as high or higher than that of the current generation.

That invites some important questions. What about equity? Most, I think, would insist that the condition of the majority of people, if not of everybody, should either stay the same or improve. And what about history? If welfare has been improving for several generations, is there a built-in expectation that historical rates of improvement will continue? Our welfare detectors, after all, are exquisitely sensitive to disparity.

Once we find agreement about what sustainability really means, we can ask what science might contribute. It is surely encouraging that science is focusing increasing attention on resource problems, but the success rate is not high. At small scales, where science is applied in limited societies where property rights can be made clear, there have been some real winners, such as managed preserves that blend conservation objectives with recreational values. But at large scales, ranging from ocean fisheries to global climate, good science often fails the implementation test because the transaction costs are too high or because political and economic factors intervene. A recommended target stock size for managing a marine fishery fails, although its stability makes it desirable, because to harvesters it looks too large to leave alone. Models and climate history tell us that global warming is likely to reach damaging levels, but the cost of controlling carbon emissions is high and there is always the mirage of a hydrogen economy, tempting us to ignore the harder present choices.

The big question in the end is not whether science can help. Plainly it could. Rather, it is whether scientific evidence can successfully overcome social, economic, and political resistance. Hardin, with his conviction that there were no technological solutions, was clearly doubtful about the prospect of a role for science. But the remaining articles in this section suggest that science may now have a central role to play.

This reformulation of the commons problem begins with Dietz, Ostrom, and Stern in the chapter "The Struggle to Govern the Commons," an overview of progress toward alternatives to the stark choice posed by Hardin. They conclude, with cautious optimism, that experience of the past 35 years has shown that paths to adaptive governance can indeed be opened, but not without a struggle. Aspects of this struggle are explored in the articles that follow.

> The big question in the end is not whether science can help. Plainly it could. Rather, it is whether scientific evidence can successfully overcome social, economic, and political resistance.

Several viewpoints discuss communication and discourse within and between institutions and disciplines concerned with the management of common resources. In "New Visions for Addressing Sustainability," McMichael et al. highlight the need for demography, economics, ecology, and epidemiology to talk to each other more effectively. Houck, in "Tales from a Troubled Marriage: Science and Law in Environmental Policy" explores the uneasy relationship between science and law in U.S. environmental policy during the decades since Hardin's article. Adams et al., in "Managing Tragedies: Understanding Conflict over Common Pool Resources" point out that disparities in the perceptions, knowledge, and beliefs of different stakeholders present barriers to effective communication between stakeholders in the management of common pool resources, and that recognition of this problem by all protagonists is an important step on the road to policymaking. The notion of social capital, which stresses the relationship between sustainability and social norms, provides a potential escape from Hardin's solutions to tragedy: Pretty, in "Social Capital and the Collective Management of Resources" provides an account of the positive outcomes that can ensue, at least at the local and regional level, when communities are able to adopt such an approach.

The remaining viewpoints examine three of the global commons: climate, food, and health. All of these are areas where the concept of social capital needs to be scaled up from local to international, and ultimately global, and where science has a partnership with policy-making. Chronic and infectious diseases are inimical to progress toward more sustainable lifestyles; Mascie-Taylor and Karim ("The Burden of Chronic Disease") discuss the scale of the global health problem and outline solutions on a scale that extends from local to global. Rosegrant and Cline, in "Global Food Security: Challenges and Policies," show that food security in the coming years depends not just on agricultural research and improvement, but on a web of other factors from education to investment in ecosystem services. Finally, "Climate Change: The Political Situation" by Watson and "The Challenge of Long-Term

Climate Change" by Hasselmann *et al.* explore the political challenges posed by global climate change, arguably the single most pressing global environmental problem.

Governing the Commons

Thomas Dietz, Elinor Ostrom, and Paul Stern have constructed an elegant response to the Hardin challenge—much of it, I suspect, encouraged by a wave of scholarship that has arisen in synchrony with the environmental movement that was barely beginning at the time Hardin wrote. Before going further, it will reward the reader to consult footnote 2, which pays respectful tribute to the enormous influence the Hardin paper has had on biologists and policymakers over the years since its publication. Elinor Ostrom, the Arthur F. Bentley Professor of Political Science at Indiana University, has played a leading role in this work, beginning with a landmark study of how Spanish irrigators craft social arrangements for allocating water. She has given us much of what is now understood about how societies make arrangements for governing the commons.

That tradition is well represented in this article's summary of what it takes to make commons governance work. There has to be information derived from monitoring; otherwise no one knows what has been taken out from the resource. Rates of change should be moderate because governance, the authors point out, is a kind of arms race in which the social arrangements must continually contend with changes in, for example, extraction technologies. It should be possible to exclude outsiders, who may be unfamiliar with the rules, at reasonably low cost. The other criteria entail what the authors call *social capital*—a term used by Ostrom in the title of a recent book, and adopted by the author of the next chapter in this one. Social capital involves, among other things, the need for members of the community to be familiar with one another, forming a social network that encourages trust and leads to a situation in which users support the rules and their enforcement.

Reaching Agreements

The examples of settings in which these arrangements work well tend to be local or at most regional. The authors emphasize that many of the most challenging environmental problems are global and complex. But they also argue that there are ways of applying some of what has been learned at smaller scale to these big systems. Some of the tools—informal communication and community-based governance—can be applied, although they have been used sparingly so far. Information about an extensive resource can be disaggregated so that it is available to local communities for constructing their own policies. The Canadian government's policies in attempting to restore the cod fishery are used by the authors as an effective illustration of the consequences of over-aggregation.

Even at the larger scales, the creation of action communities can be made to work. Trends in the development of environmental regulation have emphasized consultation with various interested parties—"stakeholders" as they are frequently called in environmental policy language. Some successes in this kind of process have been achieved in the development of water regulations even for large watersheds, such as the Mississippi in an example used by the authors. It has eased the development of some "command-and-control" regulations. For example, in a case mentioned by the authors, the EPA conducted a regulatory negotiation ("reg-neg" in environment-speak) in determining the amount of chlorinated by-products of disinfection that could be permitted in municipal water supplies. The negotiators had to balance the potential toxicity of these chlorinated compounds against the public-health values of preventing water-borne infections. Command-and-control rules, based on the "polluter pays" principle, set legal standards for some toxic air emissions, for example, or for the permissible amount of additives in processed foods. If a particular industry violates the rule, fines or even criminal penalties may result. If these rules are arrived at with careful consultation, they are likely to be respected and followed. But if they are seen as imposed arbitrarily, especially if the scientific basis for the rule is poorly understood or challenged, they may invite successful political attempts at reversal.

For that reason, that kind of regulation has sometimes been replaced by what have been called "market-based" or tradable environmental allowance rules. These developed because of a different view of pollution problems. Instead of regarding the polluters as the villains of the piece, it recognized competing interests: The polluters are a set of claimants on a common pool resource (clean water or air, for example) and the commons (everyone who uses the air or water) the other party at interest. The government, representing the commons, selects an acceptable level of utilization by the other party. Then it issues a number of permits, whose total adds up to that level; these can be auctioned or allocated in some other way.

Market-Based Regulation: An Example

The best-known success in this area, briefly considered by the authors, is the system of tradable sulfur dioxide permits mandated in the 1990 Clear Air Act Amendments. It may help the reader of this chapter to know the history of this legislation and its implementation. During the 1980s, in the Reagan era, Canada had become upset about acid rain deposition in the eastern part of that country, and forest and fishery damage was also resulting from acid rain in the United States. Tall smokestacks in Midwestern U.S. energy facilities and industries put pollutants into the air where they could be influenced by prevailing winds directed toward the east and northeast. Most of them were burning coal from Pennsylvania and West Virginia, which is relatively high in sulfur.

Growing concern about the problem caused the Environmental Protection Agency to launch the National Acid Precipitation Assessment Program, which made rather cautious estimates of economic effects. Nevertheless, there was no action until Henry Waxman (D-CA) sponsored a bill in 1986 that proposed command-and-control regulatory limits on the amount of SO_2 that could be emitted. That bill encountered opposition from industry and from Congress, and failed. Meanwhile the Canadian outrage continued, and the New England states became more concerned.

The newly elected Bush administration began work on a tradable permits plan, and Congress passed the amendments to the Clean Air Act that put the permits into play, with one allowance for each ton of SO_2 emitted. Under the plan, efficient producers, or those that switched to low sulfur coal, or those that invested in efficient scrubbers to eliminate the SO_2 could sell the permits they didn't need.

One of the requirements that the authors of this chapter emphasize is the need for information that "met high scientific standards and served the needs of decision makers and users." That criterion was clearly met in the new acid rain program: smokestack detectors for SO_2 were cheap and easy to install, and could be placed to monitor each of the significant emitters. Industry, in opposing the plan, estimated that it would be expensive, predicting that permit costs would be around $1,200. But instead, the market price for the permits—traded as commodity futures at the Chicago Board of Trade—started out at around $500. Over the next 5 years the price dropped to $140, and subsequently rose somewhat. The project was widely viewed as a success, though it was not achieved in the expected way. Relatively few power plants made capital investments in scrubbing equipment; instead they switched fuels as the cost of shipping coal by railroad from the low-S coal beds of the Powder River region of Wyoming had dropped as the result of a little-known act of railroad deregulation dating back to 1980.

It's interesting to note how many of the authors' criteria for successful trading programs were met. The stocks and flows were easily measurable and predictable: coal in, sulfur dioxide out. The number of users (in this case, the polluting entities) was modest enough to count and measure, and there was a high degree of homogeneity among that population.

The importance of social capital and building trust emerges as the centerpiece of this very thoughtful piece, which recently was named winner of the Sustainability Science Award by the Ecological Society of America. The authors close with a prescient warning that if we are to deal effectively with the relentlessly enlarging human footprint on the Earth, we will have to reach and use an understanding of large-scale commons governance quickly. The failure of the international processes to govern global climate change since 2003 constitutes an alarming footnote to that hope. Despite the transient success of the Kyoto Protocol, and despite a growing scientific consensus about the seriousness of the problem, there is not a working international agreement that includes the developing nations. More troubling still, the United States—the largest contributor of greenhouse gases—has still taken no steps to reduce the rate at which its emissions are growing.

Role of Social Capital

In the preceding chapter, the authors introduce the idea of "social capital" and point out its importance in the development of rules for governing the commons. Professor Jules Pretty, the director of the Center for Environment and Society at the University of East Anglia, is a well-known scholar specializing in the relationship between agriculture and environmental quality. His most recent book, *Agri-Culture: Reconnecting People, Land, and Nature,* points, as does his contribution here, to the nature of the networks that make up the core of social capital. The chapter begins with a table summarizing examples of groups formed for the purpose of managing various kinds of commons—watershed and catchment, irrigation, finance, forest management, and pest management. The table may have underestimated the numbers of such groups or the variety of purposes even at the time, and since it was drawn up the numbers have surely increased.

The essential idea is that the "rules of the game" for managing the resource must be embedded in a social network that entail bonds among individuals, and may be interconnected by bridging to other groups and linking up with agencies—for example, government entities. There are general rules for networks that appear to apply to a broad range of situations. The World Wide Web is such a network, and so are a variety of social networks that have been examined by scholars interested in their topology. To illustrate what has been discovered, imagine a network composed of individuals. Each individual is a node, and has connections to other individuals. But some individuals know large numbers of others, and thus have many connections; most, on the other hand, have only a few. (An example familiar to many is the "Kevin Bacon game," in which a network is made that consists of all movie actors. Each node has a link to every actor he or she has appeared with. The actor Kevin Bacon, having appeared in 50 movies, has links to a huge number of others, but most actors, of course, have appeared only once and have few links. In the entire network it is said that no actor is more than six links away from Bacon.)

In social networks, highly connected individuals have more influence than others simply because of the frequency of their contacts. Thus they have an unusually large share of the available social capital. And that, as Pretty points out, is associated with better health, higher incomes, and more education. That should apply to groups linking to other groups, just as it applies to individuals forming the groups, and it appears that the groups with exceptionally high social capital are better connected to government and to other influential institutions. This, Pretty argues, should make them more effective in developing management plans that have the support of government and other groups.

> Despite the transient success of the Kyoto Protocol, and despite a growing scientific consensus about the seriousness of [global climate change] there is not a working international agreement that includes the developing nations. More troubling still, the US–the largest contributor of greenhouse gases– has still taken no steps to reduce the rate at which its emissions are growing.

Like the authors of "The Struggle to Govern the Commons," Pretty expresses a caution about the applicability of these distributed successes at commons management to larger systems. Social capital works well when people know one another and the rules of the game are well understood. But much depends on the closed-access character of the resource. The world's larger environmental challenges relate to open-access resources—global climate, for example—and we will have to hope that social capital can be translated so it can be made to work at the international scale.

Conflict

What happens when things go wrong? William Adams and his colleagues, geographers from Cambridge and Oxford Universities, begin with an analysis of the consequence of misunderstanding the nature of a resource problem. They cite two examples of a superficial grasp of a problem: a wood fuel crisis in Asia and sub-Saharan Africa, and rangelands in east Africa carrying too many cattle. In both cases governments acted, or began actions, on the basis of assumptions that were consistent with commonly held perceptions but were proven wrong by more research or a clearly failed outcome.

That led the authors to a thoughtful generalization of the problem. Conflict among decision makers with respect to resource management may often arise because the participants arrive with preformed convictions about what the problem is. Stakeholders may arrive with personal experiences (the flood last year), access to particular bodies of research that recommend one solution or another, or with religious or moral convictions that condition their views. The authors do not ignore that in resource conflicts economic interests may be the most important basis for disagreement. But they argue that often these issues that occupy a deeper cognitive level are the main barriers to conflict resolution.

It is not only that perceptions of the problem may differ. Participants may be either aware or unaware of some past framework of decisions or rules. And they may well bring preconceptions of the validity or practicality of different solutions to the problem.

The authors are led to a conclusion that could be a significant element in dealing with such issues in the future. At the beginning of the process, it is essential that each participant put his or her own views of the situation on the table. That declaration should include a candid description of the individual's knowledge about the resource, familiarity with the background policy framework, and personal commitment to a certain set of moral or religious values that predispose toward a conclusion. Such a beginning, the authors argue, is desirable and might even be necessary. But it will not be sufficient, because economic and political power may nevertheless overdetermine the outcome.

Food Security

Among the problems confronting a growing world population, food availability has loomed large. In the early 1960s, respected scientists were forecasting widespread food crises and even famine. Although these predictions have proven to be too pessimistic, they may have been influential in restraining population growth. Today those who look at worldwide statistics of food production and population can report optimistically that things are getting better: There is more than enough food, as the gap between food production and population is widening.

Food Security and Food Availability

But there is a big difference between food availability and food security, and between aggregated data and what is happening on the ground where rural farmers live. That issue is what concerns Mark Rosegrant and Sarah Cline in "Global Food Security: Challenges and Policies." They are at the International Food Policy Research Institute in Washington, and Rosegrant has been deeply involved with the world food situation through his own institute and the Consultative Group on International Agricultural Research (CGIAR). Food security doesn't have to do with how much grain is grown in the world and how many people there are. It has to do with what can be made available to individuals—by virtue of growing it, or having the financial resources needed to purchase it or trade for it.

A middle-income American is obviously in good shape with respect to food security whether he or she has a garden or not. A poor farmer in an African or South Asian community may be all right in a year with average rainfall, but crop failure in a bad year may bring a state of malnutrition to that family's children that cannot be rectified because there is no money to buy food. Although the authors of this piece pay serious attention to aggregated world statistics, their real focus is on the ground where the people are. There, ameliorating rural poverty is linked to research and to improving water and land management through education. Education plays another role in addition to improving the productivity of the land. It also creates new opportunities for rural households by enabling men and women to engage in remunerative nonfarm work.

In fact, this role of education extends to another aspect of the food security problem. Cross-province comparisons in India have shown that women who receive more education have lower total fertility rates—an effect related also to their capacity to gain employment outside the home. A rural household with a given amount of land will have greater food security if, even without yield improvement, it has fewer children and an improved capacity to earn off-farm income.

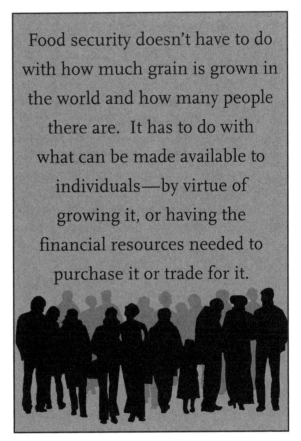

Food security doesn't have to do with how much grain is grown in the world and how many people there are. It has to do with what can be made available to individuals—by virtue of growing it, or having the financial resources needed to purchase it or trade for it.

Improving Yields

For some crops in many parts of the world, yields are no longer increasing or are increasing more slowly. Genetic research (as in the case of the improved hybrid African rice discussed by the authors) has been able to push productivity up. But the problem, as we discussed earlier in the context of Stocking's article on tropical soils, is that many small, noncommercial farmers depend on "orphan crops" rather than the major internationally traded cereal grains such as rice, wheat, and corn. Research on these has been so limited that poor rural farmers have no access to "improved" strains of whatever they depend on.

This chapter also returns us to the importance of water—often the most limiting resource in poor countries, where infrastructures for distribution and delivery are often inadequate and where, if there is competition between users, farmers often lose out. Interestingly, the authors make a case against poorly designed subsidies that encourage wasteful use—just as was described in an earlier chapter for the far richer agricultural setting in California's Central Valley. Finally, they emphasize the importance of integrating agricultural practices while paying attention to entire ecosystems and the services they provide: pollination, pest management, and the like. Local knowledge will play a role in changing how the land is managed to improve yields. So will the kinds of community-based networks that communicate that knowledge as "best practices"—yet another example of the importance of social capital.

Sustainability

These considerations take us directly to a new issue. Solutions to the problem presented by commons resources cannot be made at a moment in time and be expected to last. They must function over time, across generations—that is, be sustainable. In "New Visions for Addressing Sustainability," the Australian ecologist Anthony McMichael and his colleagues raise a set of issues that have amounted to a new coda for ecology and environmental science. A working definition of *sustainability* provided by the authors in the beginning requires that a flow of nonsubstitutable goods and services must be provided. These goods and services provide various forms of welfare for the population that depends on them. Some of these kinds of welfare—for example, "use" and "non-use"—were discussed earlier in this book. Esthetic appreciation of the natural world, for example, is one particular kind of non-use value.

Measuring and characterizing welfare are obvious tasks for those interested in sustainability. For example, suppose half the population in generation n + 1 is better off than generation n, but the other half is worse off. Even if the average amount of goods and services made available to the population as a whole had increased, the improvement would be unsatisfactory to many from an equity standpoint. Indeed, some philosophers would not accept a situation in which nearly everyone in generation n + 1 improved their welfare, as long as one was worse off. History may also play a role in terms of the human satisfactions (a form of welfare) resulting from generational improvement. Suppose that for several generations, the amount of goods and services available to a population has increased in such a way that every member has experienced improved welfare. Now a particular generation experiences no improvement, or a trivial one. Even though the technical definition of sustainability would be met by this case, people tend to make their personal judgments in terms of their own history and that of their parents. To this generation, the decrease in the experienced rate of change will be as disappointing as if there had been a loss of average welfare.

The authors' emphasis underscores the need for a broader view of sustainability. The disciplines of economics and demography are essential—but, as the authors argue, population growth is sometimes considered without clear reference to the carrying capacity of the environment. Similarly, economics based on market models sometimes fails to recognize and account adequately for the external environmental costs of actions taken within the universe of market exchange. In making the case for an interdisciplinary convergence of ecology and epidemiology with economics and demography, the authors draw attention to population-level phenomena such as culture and social conditions as they influence individual choice, and to environmental change as it affects risks to ecosystems and human health.

Chronic Disease

Sustainability as a long-term objective has to be managed in the context of threats against stability—food security, climate change, and chronic disease, for example. The burden of chronic disease is a topic for special treatment in the article by Nicholas Mascie-Taylor and Enamul Karim—biological anthropologists at Cambridge University and the Health and Life Sciences Partnership in London. Much attention in our medical past has been given to acute processes—that is, to the medical challenges posed by individual or epidemic infection. Historically, attention shifted away from the public health issues posed by the latter; it was thought that the major infectious diseases had been eliminated or drastically reduced. Medical concern, especially in the industrial countries, moved toward a focus on the diseases of older people, such as cancer, heart disease, and stroke.

But the major impacts on mortality and—especially—morbidity in the contemporary world involve chronic diseases: conditions that often pose secondary disease threats, and that, if untreated, have powerful effects on an individual's capacity to work and otherwise function in society. These dominate the global statistics of disability-adjusted life-years—the indicator that takes account of mortality plus years functionally lost because of morbidity. One often thinks of these chronic diseases as issues for the developing world, and some of them are (see below). But they are truly a global problem: Obesity, often a causative agent for diabetes and cardiovascular disease, is widespread in industrial as well as developing nations. And no population is immune to the chronic impact of HIV-AIDS or a number of vector-borne diseases.

Emerging infections—potentially epidemic diseases that were once thought to be under control, or have emerged as new agents—are a significant part of this problem, and they have received much recent attention. But in the developing world, the most significant problems are often associated with long-known parasitic infections that have not yet been dealt with effectively. Enormous effects on morbidity are associated with various parasitic soil-transmitted worms: roundworms, hookworms, and others. The total number of life-years lost from these diseases is greater than the total lost from malaria. Recognition of the extent of these problems by the World Health Assembly in 2001 has increased investments in the control of these chronic infections, but the costs of drug treatments will be heavy—and so will the costs of reducing poverty and installing adequate sanitation and other public-health infrastructures. Significant progress has been made in dealing with river-blindness (Oncocerciasis) in some places, and the control of smallpox may be followed by a similar success with polio. But several of these chronic "diseases of poverty" remain a constant reminder of the health needs of the world's poorest people.

Climate Change: Science

The problem of commons resources and how to treat them cannot be considered in the context of a static world, because the world is changing. Because the prospect of global climate change is so intimately intertwined with this and many other of the world's problems, two chapters are devoted to it in this book. The first, by a group of authors involved in the European Climate Forum, is a thoughtful and authoritative treatment of the problem as a scientific consensus now sees it. The conclusions are essentially those of the Intergovernmental Panel on Climate Change (IPCC). In talking about a scientific issue that is as red-hot politically as this one, it is advisable to start with some history. The "greenhouse" metaphor, as applied to Earth's climate, simply recognizes that Earth's atmosphere lets in much of the sun's radiation, but reflects back and thus retains much of the long-wavelength radiation responsible for heating. For more than a hundred years, it has been known that several molecules containing two kinds of atoms (carbon dioxide, water vapor, and methane) are responsible for this retention.

Over the geologic past, atmospheric concentrations of these gases have varied, and past periods of high concentration of carbon dioxide have been associated with unusually warm climates. But for the past several hundred thousands of years, CO_2 concentrations have remained at around 280 ppmv (parts per million by volume). Now, as the data presented by Hasselmann *et al.* show, the concentration of CO_2 has risen to about 380 ppmv—an increase of about 30 percent. That increase is thought by IPCC to be substantially due to human activity—the combustion of fossil fuels for energy and the burning of forests. In this chapter the authors construct models to predict what will happen in the future under various versions of a "business as usual" scenario. These projections depend on climate models that have been developed and improved at a number of climate centers over the years, and now are in good basic agreement with respect to the relationships between emissions rates and atmospheric concentrations of greenhouse gases, average global temperature, and sea level rise.

The use of "business as usual" may strike some readers as odd. The assumption is that extreme economic measures are unlikely in the short term and the resulting graphs in Figure 27 demonstrate that substantial climate effects are to be expected even if only conventional fossil fuel resources are used, and that at millennium scale these changes are extraordinary. The other graphs in the right-hand column make a set of much more optimistic assumptions about emissions mitigations and the economic development of alternatives. The dramatic difference between the two illustrates how important—especially in the long term—are the assumptions made about societal responses to the climate change problem.

Climate Change: Politics and Policy

With that we turn to the political challenge that climate change has posed for the international community, and in particular for the world's largest historical emitter of greenhouse gases, the United States. Robert Watson, before he wrote this essay, had been chair of the IPCC; he then went to the World Bank. His essay, written at the end of 2003, focuses on the so-called Kyoto Protocol. That agreement was a response to the 1992 Framework Convention on Climate Change, to which all nations—including the United States—were signatories. It declared as its goal the stabilization of greenhouse gas concentrations in the atmosphere at a level that "would prevent dangerous anthropogenic (human-caused) interference with the climate system." The Kyoto Protocol set reduction targets for carbon dioxide emissions relative to 1990 levels for participating nations, amounting to reductions of 7 to 10 percent for most nations. For the United States, it would have required substantial reductions be achieved by the period 2008–2012, but, in fact, since 1997, U.S. emissions have been on a business-as-usual trajectory. The United States has declined to ratify the Kyoto Protocol, for reasons that are well summarized and analyzed by the author.

Watson's essay is an accurate and clear summary of the political situation at the time of his writing. The Kyoto Protocol required that, in order to go into effect, nations accounting for more than 55 percent of carbon dioxide emissions must ratify it. The refusals of the United States and Australia left the swing vote with the Russian Federation, so the matter was undecided at the beginning of 2004. But Russia eventually signed, so Kyoto is in effect.

What does that mean? It is an encouraging development with little practical meaning. Its targets are not achievable by the United States even if its policies favored an attempt, which they don't. But in Europe and elsewhere, difficult emissions targets are being set and some progress is being achieved. Leading businesses (prominently British Petroleum, Shell, General Electric, and others) are announcing corporate emissions-reduction projects and are becoming politically active in favor of more aggressive government policies. At the July 2005 G-8 summit in Scotland, leaders of the industrial nations adopted a statement of principle regarding the importance of the climate change problem, but—despite strong pressure from the United Kingdom and other nations—a joint commitment to emissions reduction was not forthcoming. The open question on climate change now turns from "what will Russia do?"—where it was when Watson wrote—to "what will the United States do?"

Environmental Law and Regulation

One of the most important inheritances we have from "The Tragedy of the Commons"— surely one of the most influential scientific essays of its time—is the environmental movement and the body of laws it generated in the United States and other industrialized countries. In the United States, beginning in the mid 1960s and accelerating around 1970, the U.S. Congress passed a whole array of statutes.

Professor Oliver Houck, in "Tales from a Troubled Marriage: Science and Law in Environmental Policy," recites the alphabet soup listing these laws. Each of them mandated certain actions by administrative agencies of the federal government—most of them falling to the Environmental Protection Agency (EPA), itself a child of contemporary government reorganization.

The idea, as Houck points out, was that scientific means would be used to establish safe levels of substances or actions that were produced as a consequence of private activity. Scientifically based regulations founded on these laws would be developed by the agencies in order to eliminate or at least limit the potential "external costs" caused by private actions. Houck constructs a lively and engaging argument that in the early history of trying to develop "scientific" thresholds for various pollutants, the command-and-control regulatory process bogged down—in part at least because criteria for safety were vague in the law and unreachable by purely scientific means. More recently, regulatory provisions have tended to be couched in terms of best available technology, or schemes involving market-based mechanisms such as the sulfur dioxide permits described earlier.

But not all the things we need or want to regulate are easily handled in those ways. So, for example, the EPA must still set criteria for the National Ambient Air Quality Standards, involving such pollutants as tropospheric (street-level) ozone or small particles. Recent legal struggles over EPA proposals for new standards for both of these provide a good illustration of Houck's "cautionary tales." The American Lung Association pushed and eventually sued the EPA to force it to set the standards lower. The EPA, after internal staff work and examining new scientific literature, did so with the support of two outside scientific advisory boards. It was then taken to court by the American Trucking Association, which claimed that the EPA had failed to show a scientific "bright line" between its new standard and the Clean Air Act's criterion for safety. The circuit court found that the EPA had not made such a case, and added that it had exceeded the authority delegated to it by Congress. But the Supreme Court then reversed the circuit court's ruling and remanded the case—thus leaving the science exactly where it was, in a state of ambiguity. To most observers, it is clear that more was being asked of science than science was able to deliver. But Congress frequently sets safety standards so vaguely that agencies are tempted to use science in inappropriate ways. Houck is not against science, and argues that in many cases—climate change, for example—science can and should set the direction even when some uncertainties remain. He closes the case with an interesting story about how two distinguished scientists had a key role in conversations with a Congressional leader that rescued important environmental legislation. That tale, which is true, is unfortunately the kind of accident that is unlikely to happen often. Better, as Houck argues earlier in his essay, is that scientists "stay in the game" and work with the policy process in a systematic way.

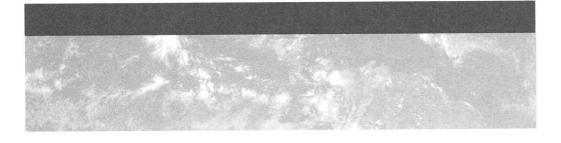

The Tragedy of the Commons

GARRETT HARDIN

The population problem has no technical solution; it requires a fundamental extension in morality.

The author is professor of biology, University of California, Santa Barbara. This article is based on a presidential address presented before the meeting of the Pacific Division of the American Association for the Advancement of Science at Utah State University, Logan, 25 June 1968.

At the end of a thoughtful article on the future of nuclear war, Wiesner and York (1) concluded that: "Both sides in the arms race are . . . confronted by the dilemma of steadily increasing military power and steadily decreasing national security. *It is our considered professional judgment that this dilemma has no technical solution.* If the great powers continue to look for solutions in the

This article first appeared in *Science* (13 December 1968: Vol. 162). It has been revised and updated for this edition.

area of science and technology only, the result will be to worsen the situation."

I would like to focus your attention not on the subject of the article (national security in a nuclear world) but on the kind of conclusion they reached, namely that there is no technical solution to the problem. An implicit and almost universal assumption of discussions published in professional and semipopular scientific journals is that the problem under discussion has a technical solution. A technical solution may be defined as one that requires a change only in the techniques of the natural sciences, demanding little or nothing in the way of change in human values or ideas of morality.

In our day (though not in earlier times) technical solutions are always welcome. Because of previous failures in prophecy, it takes courage to assert that a desired technical solution is not possible. Wiesner and York exhibited this courage; publishing in a science journal, they insisted that the solution to the problem was not to be found in the natural sciences. They cautiously qualified their statement with the phrase "It is our considered professional judgment . . ." Whether they were right or not is not the concern of the present article. Rather, the concern here is with the important concept of a class of human problems which can be called "no technical solution problems," and, more specifically, with the identification and discussion of one of these.

It is easy to show that the class is not a null class. Recall the game of tick-tack-toe. Consider the problem "How can I win the game of tick-tack-toe?" It is well known that I cannot, if I assume (in keeping with the conventions of game theory) that my opponent understands the game perfectly. Put another way, there is no "technical solution" to the problem. I can win only by giving a radical meaning to the word "win." I can hit my opponent over the head; or I can drug him; or I can falsify the records. Every way in which I "win" involves, in some sense, an abandonment of the game, as we intuitively understand it. (I can also, of course, openly abandon the game—refuse to play it. This is what most adults do.)

The class of "No technical solution problems" has members. My thesis is that the "population problem," as conventionally conceived, is a member of this class. How it is conventionally conceived needs some comment. It is fair to say that most people who anguish over the population problem are trying to find a way to avoid the evils of overpopulation without relinquishing any of the privileges they now enjoy. They think that farming the seas or developing new strains of wheat will solve the problem—technologically. I try to show here that the solution they seek cannot be found. The population problem cannot be solved in a technical way, any more than can the problem of winning the game of tick-tack-toe.

The Universal Declaration of Human Rights describes the family as the natural and fundamental unit of society. It follows that any choice and decision with regard to the size of the family must irrevocably rest with the family itself, and cannot be made by anyone else.

It is painful to have to deny categorically the validity of this right; denying it, one feels as uncomfortable as a resident of Salem, Massachusetts, who denied the reality of witches in the 17th century. At the present time, in liberal quarters, something like a taboo acts to inhibit criticism of the United Nations. There is a feeling that the United Nations is "our last and best hope," that we shouldn't find fault with it; we shouldn't play into the hands of the archconservatives. However, let us not forget what Robert Louis Stevenson said: "The truth that is suppressed by friends is the readiest weapon of the enemy." If we love the truth we must openly deny the validity of the Universal declaration of Human Rights, even though it is promoted by the United Nations. We should also join with Kingsley Davis (15) in attempting to get Planned Parenthood-World Population to see the error of its ways in embracing the same tragic ideal.

What Shall We Maximize?

Population, as Malthus said, naturally tends to grow "geometrically," or, as we would now say, exponentially. In a finite world this means that the per capita share of the world's goods must steadily decrease. Is ours a finite world?

A fair defense can be put forward for the view that the world is infinite; or that we do not know that it is not. But, in terms of the practical problems that we must face in the next few generations with the foreseeable technology, it is clear that we will greatly increase human misery if we do not, during the immediate future, assume that the world available to the terrestrial human population is finite. "Space" is no escape (2).

A finite world can support only a finite population; therefore, population growth must eventually equal zero. (The case of perpetual wide fluctuations above and below zero is a trivial variant that need not be discussed.) When this condition is met, what will be the situation of mankind? Specifically, can Bentham's goal of "the greatest good for the greatest number" be realized?

No—for two reasons, each sufficient by itself. The first is a theoretical one. It is not mathematically possible to maximize for two (or more) variables at the same time. This was clearly stated by von Neumann and Morgenstern (3), but the principle is implicit in the theory of partial differential equations, dating back at least to D'Alembert (1717–1783).

The second reason springs directly from biological facts. To live, any organism must have a source of energy (for example, food). This energy is utilized for two purposes: mere maintenance and work. For man, maintenance of life requires about 1600 kilocalories a day ("maintenance calories"). Anything that he does over and above merely staying alive will be defined as work, and is supported by "work calories" which he takes in. Work calories are used not only for what we call work in common speech; they are also required for all forms of enjoyment, from swimming and automobile racing to playing music and writing poetry. If our goal is to maximize population it is obvious what we must do: We must make the work calories per person approach as close to zero as possible. No gourmet meals, no vacations, no sports, no music, no literature, no art . . . I think that everyone will grant, without argument or proof, that maximizing population does not maximize goods. Bentham's goal is impossible.

In reaching this conclusion I have made the usual assumption that it is the acquisition of energy that is the problem. The appearance of atomic energy has led some to question this assumption. However, given an infinite source of energy, population growth still produces an inescapable problem. The problem of the acquisition of energy is replaced by the problem of its dissipation, as J. H. Fremlin has so wittily shown (4). The arithmetic signs in the analysis are, as it were, reversed; but Bentham's goal is still unobtainable.

The optimum population is, then, less than the maximum. The difficulty of defining the optimum is enormous; so far as I know, no one has seriously tackled this problem. Reaching an acceptable and stable solution will surely require more than one generation of hard analytical work—and much persuasion.

We want the maximum good per person; but what is good? To one person it is wilderness, to another it is ski lodges for thousands. To one it is estuaries to nourish ducks for hunters to shoot; to another it is factory land. Comparing one good with another is, we usually say, impossible because

> We want the maximum good per person; but what is good? To one person it is wilderness, to another it is ski lodges for thousands. To one it is estuaries to nourish ducks for hunters to shoot; to another it is factory land. Comparing one good with another is, we usually say, impossible because goods are incommensurable.

goods are incommensurable. Incommensurables cannot be compared.

Theoretically this may be true; but in real life incommensurables are commensurable. Only a criterion of judgment and a system of weighting are needed. In nature the criterion is survival. Is it better for a species to be small and hideable, or large and powerful? Natural selection commensurates the incommensurables. The compromise achieved depends on a natural weighting of the values of the variables.

Man must imitate this process. There is no doubt that in fact he already does, but unconsciously. It is when the hidden decisions are made explicit that the arguments begin. The problem for the years ahead is to work out an acceptable theory of weighting. Synergistic effects, nonlinear variation, and difficulties in discounting the future make the intellectual problem difficult, but not (in principle) insoluble.

Has any cultural group solved this practical problem at the present time, even on an intuitive level? One simple fact proves that none has: There is no prosperous population in the world today that has, and has had for some time, a growth rate of zero. Any people that has intuitively identified its optimum point will soon reach it, after which its growth rate becomes and remains zero.

Of course, a positive growth rate might be taken as evidence that a population is below its optimum. However, by any reasonable standards, the most rapidly growing populations on Earth today are (in general) the most miserable. This association (which need not be invariable) casts doubt on the optimistic assumption that the positive growth rate of a population is evidence that it has yet to reach its optimum.

We can make little progress in working toward optimum population size until we explicitly exorcize the spirit of Adam Smith in the field of practical demography. In economic affairs, *The Wealth of Nations* (1776) popularized the "invisible hand," the idea that an individual who "intends only his own gain," is, as it were, "led by an invisible hand to promote . . . the public interest" (5). Adam Smith did not assert that this was invariably true, and perhaps neither did any of his followers. But he contributed to a dominant tendency of thought that has ever since interfered with positive action based on rational analysis, namely, the tendency to assume that decisions reached individually will, in fact, be the best decisions for an entire society. If this assumption is correct it justifies the continuance of our present policy of laissez-faire in reproduction. If it is correct we can assume that men will control their individual fecundity so as to produce the optimum population. If the assumption is not correct, we need to reexamine our individual freedoms to see which ones are defensible.

Tragedy of Freedom in a Commons

The rebuttal to the invisible hand in population control is to be found in a scenario first sketched in a little-known pamphlet (6) in 1833 by a mathematical amateur named William Forster Lloyd (1794–1852). We may well call it "the tragedy of the commons," using the word "tragedy" as the philosopher Whitehead used it (7): "The essence of dramatic tragedy is not unhappiness. It resides in the solemnity of the remorseless working of things." He then goes on to say, "This inevitableness of destiny can only be illustrated in terms of human life by incidents which in fact involve unhappiness. For it is only by them that the futility of escape can be made evident in the drama."

The tragedy of the commons develops in this way. Picture a pasture open to all. It is to be expected that each herdsman will try to keep as many cattle as possible on the commons. Such an arrangement may work reasonably satisfactorily for centuries because tribal wars, poaching, and disease keep the numbers of both man and beast well below the carrying capacity of the land. Finally, however, comes the day of reckoning, that is, the day when the long-desired goal of social stability becomes a reality. At this point, the inherent logic of the commons remorselessly generates tragedy.

As a rational being, each herdsman seeks to maximize his gain. Explicitly or implicitly, more or less consciously, he asks, "What is the utility *to me* of adding one more animal to my herd?" This utility has one negative and one positive component.

1. The positive component is a function of the increment of one animal. Since the herdsman receives all the proceeds from the sale of the additional animal, the positive utility is nearly +1.

2. The negative component is a function of the additional overgrazing created by one more animal. Since, however, the effects of overgrazing are shared by all the herdsmen, the negative utility for any particular decision-making herdsman is only a fraction of −1.

Adding together the component partial utilities, the rational herdsman concludes that the only sensible course for him to pursue is to add another animal to his herd. And another; and another. . . . But this is the conclusion reached by each and every rational herdsman sharing a commons. Therein is the tragedy. Each man is locked into a system that compels him to increase his herd without limit—in a world that is limited. Ruin is the destination toward which all men rush, each pursuing his own best interest in a society that believes in the freedom of the commons. Freedom in a commons brings ruin to all.

Some would say that this is a platitude. Would that it were! In a sense, it was learned thousands of years ago, but natural selection favors the forces of psychological denial (8). The individual benefits as an individual from his ability to deny the truth even though society as a whole, of which he is a part, suffers.

Education can counteract the natural tendency to do the wrong thing, but the inexorable succession of generations requires that the basis for this knowledge be constantly refreshed.

A simple incident that occurred a few years ago in Leominster, Massachusetts, shows how perishable the knowledge is. During the Christmas shopping season the parking meters downtown were covered with plastic bags that bore tags reading: "Do not open until after Christmas. Free parking courtesy of the mayor and city council." In other words, facing the prospect of an increased demand for already scarce space, the city fathers reinstituted the system of the commons. (Cynically, we suspect that they gained more votes than they lost by this retrogressive act.)

In an approximate way, the logic of the commons has been understood for a long time, perhaps since the discovery of agriculture or the invention of private property in real estate. But it is understood mostly only in special cases which are not sufficiently generalized. Even at this late date, cattlemen leasing national land on the western ranges demonstrate no more than an ambivalent understanding, in constantly pressuring federal authorities to increase the head count to the point where overgrazing produces erosion and weed-dominance. Likewise, the oceans of the world continue to suffer from the survival of the philosophy of the commons. Maritime nations still respond automatically to the shibboleth of the "freedom of the seas." Professing to believe in the "inexhaustible resources of the oceans," they bring species after species of fish and whales closer to extinction (9).

The national parks present another instance of the working out of the tragedy of the commons. At present, they are open to all, without limit. The parks themselves are limited in extent—there is only one Yosemite Valley—whereas population seems to grow without limit. The values that visitors seek in the parks are steadily eroded. Plainly, we must soon cease to treat the parks as commons or they will be of no value to anyone.

What shall we do? We have several options. We might sell them off as private property. We might keep them as public property, but allocate the right to enter them. The allocation might be on the basis of wealth, by the use of an auction system. It might be on the basis of merit, as defined by some agreed-upon standards. It might be by lottery. Or it might be on a first-come, first-served basis, administered to long queues. These, I think, are all the reasonable possibilities. They are all objectionable. But we must choose—or acquiesce in the destruction of the commons that we call our national parks.

Pollution

In a reverse way, the tragedy of the commons reappears in problems of pollution. Here it is not a question of taking something out of the commons, but of putting something in—sewage, or chemical, radioactive, and heat wastes into water; noxious and dangerous fumes into the air; and distracting and unpleasant advertising signs into the line of sight. The calculations of utility are much the same as before. The rational man finds that his share of the cost of the wastes he discharges into the commons is less than the cost of purifying his wastes before releasing them. Since this is true for everyone, we are locked into a system of "fouling our own nest," so long as we behave only as independent, rational, free-enterprisers.

The tragedy of the commons as a food basket is averted by private property, or something formally like it. But the air and waters surrounding us cannot readily be fenced, and so the tragedy of the commons as a cesspool must be prevented by different means, by coercive laws or taxing devices that make it cheaper for the polluter to treat his pollutants than to discharge them untreated. We have not progressed as far with the solution of this problem as we have with the first. Indeed, our particular concept of private property, which deters us from exhausting the positive resources of the earth, favors pollution. The owner of a factory on the bank of a stream—whose property extends to the middle of the stream—often has difficulty seeing why it is not his natural right to muddy the waters flowing past his door. The law, always behind the times, requires elaborate stitching and fitting to adapt it to this newly perceived aspect of the commons.

The pollution problem is a consequence of population. It did not much matter how a lonely American frontiersman disposed of his waste. "Flowing water purifies itself every 10 miles," my grandfather used to say, and the myth was near enough to the truth when he was a boy, for there were not too many people. But as population became denser, the natural chemical and biological recycling processes became overloaded, calling for a redefinition of property rights.

How to Legislate Temperance?

Analysis of the pollution problem as a function of population density uncovers a not generally recognized principle of morality, namely: *the morality of an act is a function of the state of the system at the time it is performed* (10). Using the commons as a cesspool does not harm the general public under frontier conditions, because there is no public; the same behavior in a metropolis is unbearable. A hundred and fifty years ago a plainsman could kill an American bison, cut out

only the tongue for his dinner, and discard the rest of the animal. He was not in any important sense being wasteful. Today, with only a few thousand bison left, we would be appalled at such behavior.

In passing, it is worth noting that the morality of an act cannot be determined from a photograph. One does not know whether a man killing an elephant or setting fire to the grassland is harming others until one knows the total system in which his act appears. "One picture is worth a thousand words," said an ancient Chinese; but it may take 10,000 words to validate it. It is as tempting to ecologists as it is to reformers in general to try to persuade others by way of the photographic shortcut. But the essense of an argument cannot be photographed: it must be presented rationally—in words.

That morality is system-sensitive escaped the attention of most codifiers of ethics in the past. "Thou shalt not . . ." is the form of traditional ethical directives which make no allowance for particular circumstances. The laws of our society follow the pattern of ancient ethics, and therefore are poorly suited to governing a complex, crowded, changeable world. Our epicyclic solution is to augment statutory law with administrative law. Since it is practically impossible to spell out all the conditions under which it is safe to burn trash in the backyard or to run an automobile without smog-control, by law we delegate the details to bureaus. The result is administrative law, which is rightly feared for an ancient reason—*Quis custodiet ipsos custodes?*— "Who shall watch the watchers themselves?" John Adams said that we must have "a government of laws and not men." Bureau administrators, trying to evaluate the morality of acts in the total system, are singularly liable to corruption, producing a government by men, not laws.

Prohibition is easy to legislate (though not necessarily to enforce); but how do we legislate temperance? Experience indicates that it can be accomplished best through the mediation of administrative law. We limit possibilities unnecessarily if we suppose that the sentiment of *Quis custodiet* denies us the use of administrative law. We should rather retain the phrase as a perpetual reminder of fearful dangers we cannot avoid. The great challenge facing us now is to invent the corrective feedbacks that are needed to keep custodians honest. We must find ways to legitimate the needed authority of both the custodians and the corrective feedbacks.

Freedom to Breed Is Intolerable

The tragedy of the commons is involved in population problems in another way. In a world governed solely by the principle of "dog eat dog" — if indeed there ever was such a world—how many children a family had would not be a matter of public concern. Parents who bred too exuberantly would leave fewer descendants, not more, because they would be unable to care adequately for their children. David Lack and others have found that such a negative feedback demonstrably controls the fecundity of birds (11). But men are not birds, and have not acted like them for millenniums, at least.

If each human family were dependent only on its own resources; if the children of improvident parents starved to death; *if*, thus, overbreeding brought its own "punishment" to the germ line— *then* there would be no public interest in controlling the breeding of families. But our society is deeply committed to the welfare state (12), and hence is confronted with another aspect of the tragedy of the commons.

In a welfare state, how shall we deal with the family, the religion, the race, or the class (or indeed any distinguishable and cohesive group) that adopts overbreeding as a policy to secure its own aggrandizement (13)? To couple the concept of freedom to breed with the belief that everyone born has an equal right to the commons is to lock the world into a tragic course of action.

Unfortunately this is just the course of action that is being pursued by the United Nations. In late 1967, some 30 nations agreed to the following (14): The Universal Declaration of Human Rights describes the family as the natural and fundamental unit of society. It follows that any choice and decision with regard to the size of the family must irrevocably rest with the family itself, and cannot be made by anyone else.

It is painful to have to deny categorically the validity of this right; denying it, one feels as

In a world governed solely by the principle of "dog eat dog"— if indeed there ever was such a world—how many children a family had would not be a matter of public concern. Parents who bred too exuberantly would leave fewer descendants, not more, because they would be unable to care adequately for their children.

uncomfortable as a resident of Salem, Massachusetts, who denied the reality of witches in the 17th century. At the present time, in liberal quarter, something like a taboo acts to inhibit criticism of the United States. There is a feeling that the United States is "our last and best hope," that we shouldn't find fault with it; we shouldn't play into the hands of archconservatives. However, let us not forget what Robert Louis Stevenson said: "The truth that is suppressed by friends is the readiest weapon of the enemy." If we love the truth, we must openly deny the validity of the Universal Declaration of Human Rights, even though it is promoted by the United Nations. We should also join with Kingsley Davis (15) in attempting to get Planned Parenthood-World Population to see the error of its ways in embracing the same tragic ideal.

Conscience Is Self-Eliminating

It is a mistake to think that we can control the breeding of mankind in the long run by an appeal to conscience. Charles Galton Darwin made this point when he spoke on the centennial of the publication of his grandfather's great book. The argument is straightforward and Darwinian.

People vary. Confronted with appeals to limit breeding, some people will undoubtedly respond to the plea more than others. Those who have more children will produce a larger fraction of the next generation than those with more susceptible consciences. The difference will be accentuated, generation by generation.

In C. G. Darwin's words: "It may well be that it would take hundreds of generations for the progenitive instinct to develop in this way, but if it should do so, nature would have taken her revenge, and the variety *Homo contracipiens* would become extinct and would be replaced by the variety *Homo progenitivus*" (16).

The argument assumes that conscience or the desire for children (no matter which) is hereditary—but hereditary only in the most general formal sense. The result will be the same whether the attitude is transmitted through germ cells, or exosomatically, to use A. J. Lotka's term. (If one denies the latter possibility as well as the former, then what's the point of education?) The argument has here been stated in the context of the population problem, but it applies equally well to any instance in which society appeals to an individual exploiting a commons to restrain himself for the general good—by means of his conscience. To make such an appeal is to set up a selective system that works toward the elimination of conscience from the race.

Pathogenic Effects of Conscience

The long-term disadvantage of an appeal to conscience should be enough to condemn it but has serious short-term disadvantages as well. If we ask a man who is exploiting a commons to desist "in the name of conscience," what are we saying to him? What does he hear?—not only at the moment but also in the wee small hours of the night when, half asleep, he remembers not merely the words we used but also the nonverbal communication cues we gave him unawares? Sooner or later, consciously or subconsciously, he senses that he has received two communications, and that they are contradictory: (i) (intended communication) "If you don't do as we ask, we will openly condemn you for not acting like a responsible citizen"; (ii) (the unintended communication) "If you do behave as we ask, we will secretly condemn you for a simpleton who can be shamed into standing aside while the rest of us exploit the commons."

Everyman then is caught in what Bateson has called a "double bind." Bateson and his co-workers have made a plausible case for viewing the double bind as an important causative factor in the genesis of schizophrenia (17). The double bind may not always be so damaging, but it always endangers the mental health of anyone to whom it is applied. "A bad conscience," said Nietzsche, "is a kind of illness."

To conjure up a conscience in others is tempting to anyone who wishes to extend his control beyond the legal limits. Leaders at the highest level succumb to this temptation. Has any President during the past generation failed to call on labor unions to moderate voluntarily their demands for higher wages, or to steel companies to honor voluntary guidelines on prices? I can recall none. The rhetoric used on such occasions is designed to produce feelings of guilt in noncooperators.

For centuries it was assumed without proof that guilt was a valuable, perhaps even an indispensable, ingredient of the civilized life. Now, in this post-Freudian world, we doubt it.

Paul Goodman speaks from the modern point of view when he says: "No good has ever come from feeling guilty, neither intelligence, policy, nor compassion. The guilty do not pay attention to the object but only to themselves, and not even to their own interests, which might make sense, but to their anxieties" (18).

One does not have to be a professional psychiatrist to see the consequences of anxiety. We in the Western world are just emerging from a dreadful two-centuries-long Dark Ages of Eros that was sustained partly by prohibition laws, but perhaps more effectively by the anxiety-generating mechanism of education. Alex Comfort has told the story well in *The Anxiety Makers* (19); it is not a pretty one.

Since proof is difficult, we may even concede that the results of anxiety may sometimes, from certain points of view, be desirable. The larger question we should ask is whether, as a matter of policy, we should ever encourage the use of a technique the tendency (if not the intention) of which is psychologically pathogenic. We hear much talk these days of responsible parenthood; the coupled words are incorporated into the titles of some organizations devoted to birth control. Some people have proposed massive propaganda campaigns to instill responsibility into the nation's (or the world's) breeders. But what is the meaning of the word responsibility in this context? Is it not merely a synonym for the word conscience? When we use the word responsibility in the absence of substantial sanctions are we not trying to browbeat a free man in a commons into acting against his own interest? Responsibility is a verbal counterfeit for a substantial *quid pro quo*. It is an attempt to get something for nothing.

If the word responsibility is to be used at all, I suggest that it be in the sense Charles Frankel uses it (20). "Responsibility," says this philosopher, "is the product of definite social arrangements." Notice that Frankel calls for social arrangements—not propaganda.

Mutual Coercion Mutually Agreed Upon

The social arrangements that produce responsibility are arrangements that create coercion, of some sort. Consider bank-robbing. The man who takes money from a bank acts as if the bank were a commons. How do we prevent such action? Certainly not by trying to control his behavior solely by a verbal appeal to his sense of responsibility. Rather than rely on propaganda we follow Frankel's lead and insist that a bank is not a commons; we seek the definite social arrangements that will keep it from becoming a commons. That we thereby infringe on the freedom of would-be robbers we neither deny nor regret.

The morality of bank-robbing is particularly easy to understand because we accept complete prohibition of this activity. We are willing to say "Thou shalt not rob banks," without providing for exceptions. But temperance also can be created by coercion. Taxing is a good coercive device. To keep downtown shoppers temperate in their use of parking space we introduce parking meters for short periods, and traffic fines for longer ones. We need not actually forbid a citizen to park as long as he wants to; we need merely make it increasingly expensive for him to do so. Not prohibition, but carefully biased options are what we offer him. A Madison Avenue man might call this persuasion; I prefer the greater candor of the word coercion.

Coercion is a dirty word to most liberals now, but it need not forever be so. As with the four-letter words, its dirtiness can be cleansed away by exposure to the light, by saying it over and over without apology or embarrassment. To many, the word coercion implies arbitrary decisions of distant and irresponsible bureaucrats; but this is not a necessary part of its meaning. The only kind of coercion I recommend is mutual coercion, mutually agreed upon by the majority of the people affected.

To say that we mutually agree to coercion is not to say that we are required to enjoy it, or even to pretend we enjoy it. Who enjoys taxes? We all grumble about them. But we accept compulsory taxes because we recognize that voluntary taxes would favor the conscienceless. We institute and (grumblingly) support taxes and other coercive devices to escape the horror of the commons.

An alternative to the commons need not be perfectly just to be preferable. With real estate and other material goods, the alternative we have chosen is the institution of private property coupled with legal inheritance. Is this system perfectly just? As a genetically trained biologist I deny that it is. It seems to me that, if there are to be differences in individual inheritance, legal possession should be perfectly correlated with biological inheritance—that those who are biologically more fit to be the custodians of property and power should legally inherit more. But genetic recombination continually makes a mockery of the doctrine of "like father, like son" implicit in our laws of legal inheritance. An idiot can inherit millions, and a trust fund can keep his estate intact. We must admit that our legal system of private property plus inheritance is unjust—but we put up with it because we are not convinced, at the moment, that anyone has invented a better system. The alternative of the commons is too horrifying to contemplate. Injustice is preferable to total ruin.

It is one of the peculiarities of the warfare between reform and the status quo that it is thoughtlessly governed by a double standard. Whenever a reform measure is proposed it is often defeated when its opponents triumphantly discover a flaw in it. As Kingsley Davis has pointed out (*21*), worshippers of the status quo sometimes imply that no reform is possible without unanimous agreement, an implication contrary to historical fact. As nearly as I can make out, automatic rejection of proposed reforms is based on one of two unconscious assumptions: (i) that the status quo is perfect; or (ii) that the choice we face is between reform and no action; if the proposed reform is imperfect, we presumably should take no action at all, while we wait for a perfect proposal.

But we can never do nothing. That which we have done for thousands of years is also action. It also produces evils. Once we are aware that the status quo is action, we can then compare its discoverable advantages and disadvantages with the predicted advantages and disadvantages of the proposed reform, discounting as best we can for our lack of experience. On the basis of such a comparison, we can make a rational decision which will not involve the unworkable assumption that only perfect systems are tolerable.

Recognition of Necessity

Perhaps the simplest summary of this analysis of man's population problems is this: The commons, if justifiable at all, is justifiable only under conditions of low-population density. As the human population has increased, the commons has had to be abandoned in one aspect after another.

First we abandoned the commons in food gathering, enclosing farm land and restricting pastures and hunting and fishing areas. These restrictions are still not complete throughout the world.

Somewhat later we saw that the commons as a place for waste disposal would also have to be abandoned. Restrictions on the disposal of domestic sewage are widely accepted in the Western world; we are still struggling to close the commons to pollution by automobiles, factories, insecticide sprayers, fertilizing operations, and atomic energy installations.

In a still more embryonic state is our recognition of the evils of the commons in matters of pleasure. There is almost no restriction on the propagation of sound waves in the public medium. The shopping public is assaulted with mindless music, without its consent. Our government is paying out billions of dollars to create supersonic transport which will disturb 50,000 people for every one person who is whisked from coast to coast 3 hours faster. Advertisers muddy the airwaves of radio and television and pollute the view of travelers. We are a long way from outlawing the commons in matters of pleasure. Is this because our Puritan inheritance makes us view pleasure as something of a sin, and pain (that is, the pollution of advertising) as the sign of virtue?

Every new enclosure of the commons involves the infringement of somebody's personal liberty. Infringements made in the distant past are accepted because no contemporary complains of a loss. It is the newly proposed infringements that we vigorously oppose; cries of "rights" and "freedom" fill the air. But what does "freedom" mean? When men mutually agreed to pass laws against robbing, mankind became more free, not less so. Individuals locked into the logic of the commons are free only to bring on universal ruin; once they see the necessity of mutual coercion, they become free to pursue other goals. I believe it was Hegel who said, "Freedom is the recognition of necessity."

The most important aspect of necessity that we must now recognize, is the necessity of abandoning the commons in breeding. No technical solution can rescue us from the misery of overpopulation. Freedom to breed will bring ruin to all. At the moment, to avoid hard decisions many of us are tempted to propagandize for conscience and responsible parenthood. The temptation must be resisted, because an appeal to independently acting consciences selects for the disappearance of all conscience in the long run, and an increase in anxiety in the short.

The only way we can preserve and nurture other and more precious freedoms is by relinquishing the freedom to breed, and that very soon. "Freedom is the recognition of necessity" — and it is the role of education to reveal to all the necessity of abandoning the freedom to breed. Only so, can we put an end to this aspect of the tragedy of the commons.

References

1. J. B. Wiesner and H. F. York, *Sci. Amer.* 211 (No. 4), 27 (1964).
2. G. Hardin, *J. Hered.* 50, 68 (1959) ; S. von Hoernor, *Science* 137, 18 (1962) .
3. J. von Neumann and O. Morgenstern, *Theory of Games and Economic Behavior* (Princeton Univ. Press, Princeton, N.J., 1947), p.11.
4. J. H. Fremlin, *New Sci.*, No. 415 (1964), p. 285.
5. A. Smith, *The Wealth of Nations* (Modern Library, New York, 1937), p. 423.
6. W. F. Lloyd, *Two Lectures on the Checks to Population* (Oxford Univ. Press, Oxford, England, 1833), reprinted (in part) in *Population, Evolution, and Birth Control*, G. Hardin, Ed. (Freeman, San Francisco, 1964), p. 37.
7. A. N. Whitehead, *Science and the Modern World* (Mentor, New York, 1948), p. 17.
8. G. Hardin, Ed. *Population, Evolution, and Birth Control* (Freeman, San Francisco, 1964), p. 56.
9. S. McVay, *Sci. Amer.* 216 (No. 8), 13 (1966).
10. J. Fletcher, *Situation Ethics* (Westminster, Philadelphia, 1966).
11. D. Lack, *The Natural Regulation of Animal Numbers* (Clarendon Press, Oxford, 1954).
12. H. Girvetz, *From Wealth to Welfare* (Stanford Univ. Press, Stanford, Calif., 1950).
13. G. Hardin, *Perspec. Biol. Med.* 6, 366 (1963).
14. U. Thant, *Int. Planned Parenthood News*, No. 168 (February 1968), p. 3.
15. K. Davis, *Science* 158, 730 (1967) .
16. S. Tax, Ed., *Evolution after Darwin* (Univ. of Chicago Press, Chicago, 1960), vol. 2, p. 469.
17. G. Bateson, D. D. Jackson, J. Haley, J. Weakland, *Behav. Sci.* 1, 251 (1956) .
18. P. Goodman, *New York Rev. Books* 10(8), 22 (23 May 1968).
19. A. Comfort, *The Anxiety Makers* (Nelson, London, 1967).
20. C. Frankel, *The Case for Modern Man* (Harper, New York, 1955), p. 203.
21. J. D. Roslansky, *Genetics and the Future of Man* (Appleton-Century-Crofts, New York, 1966), p. 177.

The Struggle to Govern the Commons

THOMAS DIETZ, ELINOR OSTROM,
AND PAUL C. STERN

I n 1968, Garrett Hardin (1) drew attention to two human factors that drive environmental change. The first factor is the increasing demand for natural resources and environmental services, stemming from growth in human population and per capita resource consumption. The second factor is the way in which humans organize themselves to extract resources from the environment and eject effluents into it—what social scientists refer to as institutional arrangements. Hardin's work has been highly influential (2) but has long been aptly criticized as oversimplified (3–6).

Hardin's oversimplification was twofold: He claimed that only two state-established institutional arrangements—centralized government and private property—could sustain commons over the long run, and he presumed that resource users were trapped in a commons dilemma,

This article first appeared in *Science* (12 December 2003: Vol. 302, no. 5652). It has been revised and updated for this edition.

unable to create solutions (7–9). He missed the point that many social groups, including the herders on the commons that provided the metaphor for his analysis, have struggled suc-

KEY TERM

Inshore fisheries are usually defined as capture fisheries within 12 nautical miles of the coast. The use of the term reflects both legal and cultural traditions, as well as the nature of fishing grounds and the structure of national fleets. For example, in England, **inshore fisheries** are often defined as fisheries worked by vessels under 30 feet (10 meters) typically fishing out to 6 miles, under an inshore management regime.

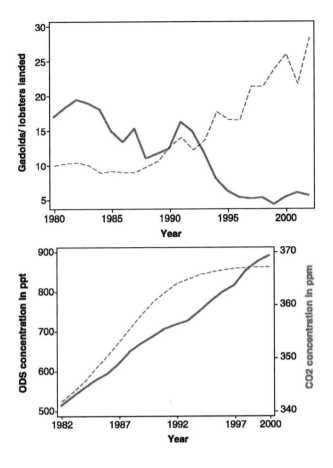

FIGURE 21. Comparison of landings of groundfish (gadoids, solid line) and lobster (dashed line) in Maine from 1980 to 2002, measured in millions of kilograms of groundfish and lobsters landed per year. International fishing in these waters ended with the extended jurisdiction that occurred in 1977 (167).

FIGURE 22. Atmospheric concentration of CO_2 (solid line, right scale) and three principal ODS (dashed line, left scale). The ODS are chlorofluorocarbons (CFCs) 11, 12, and 113 and were weighted based on their ozone-depleting potential (168). Data are from (169). *ppt* = parts per trillion; *ppm* = parts per million.

cessfully against threats of resource degradation by developing and maintaining self-governing institutions (3, 10–14). Although these institutions have not always succeeded, neither have Hardin's preferred alternatives of private or state ownership.

In the absence of effective governance institutions at the appropriate scale, natural resources and the environment are in peril from increasing human population, consumption, and deployment of advanced technologies for resource use, all of which have reached unprecedented levels. For example, it is estimated that "the global ocean has lost more than 90% of large predatory fishes" with an 80 percent decline typically occurring "within 15 years of industrialized exploitation" (15). The threat of massive ecosystem degradation results from an interplay among ocean ecologies, fishing technologies, and inadequate governance.

Inshore fisheries are similarly degraded where they are open access or governed by top-

down national regimes, leaving local and regional officials and users without sufficient autonomy and understanding to design effective institutions (16, 17). For example, the degraded inshore ground fishery in Maine is governed by top-down rules based on models that were not credible to users. As a result, compliance has been relatively low and there has been strong resistance to strengthening existing restrictions. This is in marked contrast to the Maine lobster fishery, which has been governed by formal and informal user institutions that have strongly influenced state-level rules that restrict fishing. The result has been credible rules with very high levels of compliance (18–20). A comparison of the landings of groundfish and lobster since 1980 is shown in Figure 21. The rules and high levels of compliance related to lobster appear to have prevented the destruction of this fishery but probably are not responsible for the sharp rise in abundance and landings after 1986.

Resources at broader scales have also been successfully protected through appropriate international governance regimes such as the **Montreal Protocol** on stratospheric ozone and the International Commission for the Protection of the Rhine Agreements (*21–25*). Figure 22 compares the trajectory of atmospheric concentrations of ozone-depleting substances (ODS) with that of carbon dioxide since 1982. The Montreal Protocol, the centerpiece of the international agreements on ozone depletion, was signed in 1987. Before then, ODS concentrations were

---

KEY TERM

The "**Montreal Protocol** on Substances That Deplete the Ozone Layer" was originally signed in 1987 and substantially amended in 1990 and 1992. It originally stipulated that the production and consumption of chlorofluorocarbons (CFCs), halons, carbon tetrachloride, and methyl chloroform were to be phased out by 2000 (2005 for methyl chloroform). As of September 2002, 183 countries had ratified the Montreal Protocol. Production and consumption of ozone-depleting chemicals have been phased out in industrialized countries, and a schedule is in place to eliminate the use of methyl bromide, a pesticide and agricultural fumigant. Developing countries agreed to reduce consumption by 50 percent by January 1, 2005, and to eliminate CFCs by January 1, 2010.

---

increasing faster than those of CO_2; the increases slowed by the early 1990s and the concentration appears to have stabilized in recent years. The international treaty regime to reduce the anthropogenic impact on stratospheric ozone is widely considered an example of a successful effort to protect the global commons. In contrast, international efforts to reduce greenhouse gas concentrations have not yet had an impact.

Knowledge from an emerging science of human-environment interactions, sometimes called human ecology or the "second environmental science" (*26, 27*), is revealing which characteristics of institutions facilitate and which undermine sustainable use of environmental resources under particular conditions (*6, 28*). We know most about small-scale ecologies and institutions whose many successes and failures have been studied for years. Researchers are now developing a knowledge base for broader-scale systems. In this review, we address what science has learned about governing the commons, how adaptive governance can be implemented, and why it is always a struggle (*29*).

Why a Struggle?

Devising ways to sustain the Earth's ability to support diverse life, including a reasonable quality of life for humans, involves making tough decisions under uncertainty, complexity, and substantial biophysical constraints as well as conflicting human values and interests. Devising effective governance systems is akin to a coevolutionary race. A set of rules crafted to fit one set of socioecological conditions can erode as social, economic, and technological developments increase the potential for human damage to ecosystems and even to the biosphere itself. Furthermore, humans devise ways of evading governance rules. Thus, successful commons governance requires that rules evolve.

Effective commons governance is easier to achieve when (i) the resources and use of the resources by humans can be monitored, and the information can be verified and understood at rel-

SCIENCE IN THE NEWS

Science, Vol. 308, no. 5724, 937, 13 May 2005

FISH MOVED BY WARMING WATERS
Mason Inman

Climate change has fish populations on the move. In Europe's intensively fished North Sea, the warming waters over the past quarter-century have driven fish populations northward and deeper, according to a study by conservation ecologist John D. Reynolds of the University of East Anglia in Norwich, U.K., and his colleagues. Such warming could hamper the revival of overfished species and disrupt ecosystems, they assert. The warming is expected to continue in the North Sea, and although fish species living to the south will likely move north and replace departing ones, the forecast for the region's fisheries will depend on whether the species that succeed are marketable.

Gone fish. Warming waters in the North Sea may make it harder for commercial fishers to find their normal catch.
CREDIT: DIRK FREDER/ISTOCKPHOTO.COM

"This is another clear indication that warming is playing a role" in ocean ecosystems, says physical oceanographer Ken Drinkwater of the Institute of Marine Research in Bergen, Norway. Although there have been many studies looking at the effects of climate change on marine species, "no one has looked in detail at changes in distributions of commercial and noncommercial species," says fish biologist Paul Hart of the University of Leicester in the United Kingdom. Similar climate-induced shifts in fish populations, he adds, might happen in other temperate seas, including those around Europe and much of the United States.

The study used extensive records of fishing catches made by research vessels between 1977 and 2001, a period during which the North Sea's waters warmed by 1°C at the sea floor. Reynolds's team cast a wide net, compiling data on the sea's 36 most common bottom-dwelling fish. They found that two-thirds of the populations moved toward cooler waters—either going north or to deeper waters, or both. "We saw shifts in both commercial and noncommercial species, and across a broad set of species," says conservation ecologist Allison Perry of the University of East Anglia. The fish species whose distribution have shifted tend to be smaller and mature earlier, she and her colleagues noted.

atively low cost (e.g., trees are easier to monitor than fish, and lakes are easier to monitor than rivers) (30); (ii) rates of change in resources, resource-user populations, technology, and economic and social conditions are moderate (31–33);

(iii) communities maintain frequent face-to-face communication and dense social networks—sometimes called social capital—that increase the potential for trust, allow people to express and see emotional reactions to distrust, and lower the cost of monitoring behavior and inducing rule compliance (34–37); (iv) outsiders can be excluded at relatively low cost from using the resource (new entrants add to the harvesting pressure and typically lack understanding of the rules); and (v) users support effective monitoring and rule enforcement (38–40). Few settings in the world are characterized by all of these conditions. The challenge is to devise institutional arrangements that help to establish such conditions or, as we discuss below, meet the main challenges of governance in the absence of ideal conditions (6, 41, 42).

Selective Pressures

Many subsistence societies present favorable conditions for the evolution of effective self-governing resource institutions (13). There are hundreds of documented examples of long-term sustainable resource use in such communities as well as in more economically advanced communities with effective, local, self-governing rights, but there are also many failures (6, 11, 43–45). As human communities have expanded, the selective pressures on environmental governance institutions increasingly have come from outside influences. Commerce has become regional, national, and global, and institutions at all of these levels have been created to enable and regulate trade, transportation, competition, and conflict (46, 47). These institutions shape environmental impact, even if they are not designed with that intent. They also provide mechanisms for environmental governance (e.g., national laws) and part of the social context for local efforts at environmental governance. Broader-scale governance may authorize local control, help it, hinder it, or override it (48–53). Now, every local place is strongly influenced by global dynamics (49, 54–58).

The most important contemporary environmental challenges involve systems that are intrinsically global (e.g., climate change) or are tightly linked to global pressures (e.g., timber production for the world market) and that require governance at levels from the global all the way down to the local (49, 59, 60). These situations often feature environmental outcomes spatially distant from their causes and hard-to-monitor, broader-scale economic incentives that may not be closely aligned with the condition of local ecosystems. Also, differences in power within user groups or across scales allow some to ignore rules of commons use or to reshape the rules in their own interest, such as when global markets reshape demand for local resources (e.g., forests) in ways that swamp the ability of locally evolved institutions to regulate their use (61–63).

The store of governance tools and ways to modify and combine them is far greater than often is recognized (6, 64–66). Global and national environmental policy frequently ignores community-based governance and traditional tools, such as informal communication and sanctioning, but these tools can have significant impact (64, 67). Further, no single, broad type of ownership—government, private, or community—uniformly succeeds or fails to halt major resource deterioration, as shown for forests in multiple countries (68).

Requirements of Adaptive Governance in Complex Systems

Providing information. Environmental governance depends on good, trustworthy information about stocks, flows, and processes within the resource systems being governed, as well as about the human-environment interactions that affect those systems. This information must be congruent in scale with environmental events and decisions (49, 69). Highly aggregated information may ignore or average out local information that is important in identifying future problems and developing solutions.

For example, in 2002, a moratorium on all fishing for northern cod was declared by the Canadian government after a collapse of this valuable fishery. An earlier near-collapse had led Canada to declare a 200-mile zone of exclusive fisheries jurisdiction in 1977 (70, 71). There was considerable optimism during the 1980s that the stocks, as estimated by fishery scientists, were rebuilding. Consequently, generous total catch limits were established for northern cod and other groundfish, the number of licensed fishers was allowed to increase considerably, and substantial government subsidies were allocated for new vessels (72). What went wrong? There were a variety of information-related problems, including that fisheries managers (i) treated all northern cod as a single stock instead of recognizing distinct populations with different characteristics, (ii) ignored the variability of year classes of northern cod, (iii) focused on offshore-fishery landing data rather than inshore data to "tune" the stock assessment, and (iv) ignored inshore fishers who were catching ever-smaller fish and doubted the validity of stock assessments (72–74). This experience illustrates the need to collect and model both local and aggregated information about resource conditions and to use it in making policy at the appropriate scales.

Information also must be congruent with decision makers' needs in terms of timing, content, and form of presentation (75–77). Informational systems that simultaneously meet high scientific standards and serve ongoing needs of decision makers and users are particularly useful. Information must not overload the capacity of users to assimilate it. Systems that adequately characterize environmental conditions or human activities with summary indicators—such as prices for products or emission permits, or certification of good environmental performance—can provide valuable signals as long as they are attentive to local as well as aggregate conditions (78–80).

Effective governance requires not only factual information about the state of the environment and human actions but also information about uncertainty and values. Scientific understanding of coupled human-biophysical systems will always be uncertain because of inherent unpredictability in the systems and because the science is never complete (81). Decision makers need information that characterizes the types and magnitudes of this uncertainty, as well as the nature and

extent of scientific ignorance and disagreement (82). Also, because every environmental decision requires tradeoffs, knowledge is needed about individual and social values and about the effects of decisions on various valued outcomes. For many environmental systems, local and easily cap-

> Environmental governance depends on good, trustworthy information about stocks, flows, and processes within the resource systems being governed, as well as about how human-environment interactions affect those systems.

tured values (e.g., the market value of lumber) have to be balanced against global, diffuse, and hard-to-capture values (e.g., biodiversity and the capability of humans and ecosystems to adapt to unexpected events). Finding ways to measure and monitor the outcomes for such varied values in the face of globalization is a major informational challenge for governance.

Dealing with conflict. Sharp differences in power and in values across interested parties make conflict inherent in environmental choices. Indeed, conflict resolution may be as important a motivation for designing resource institutions as is concern with the resources themselves (83). People bring varying perspectives, interests, and fundamental philosophies to problems of environmental governance (76, 84–86); their conflicts, if they do not escalate to the point of dysfunction, can spark learning and change (87, 88).

For example, a broadly participatory process was used to examine alternative strategies for regulating the Mississippi River and its tributaries

(89). A dynamic model was constructed with continuous input by the Corps of Engineers, the Fish and Wildlife Service, local landowners, environmental groups, and academics from multiple disciplines. After extensive model development and testing against past historical data, most stakeholders had high confidence in the explanatory power of the model. Consensus was reached over alternative governance options, and the resulting policies generated far less conflict than had existed at the outset (90).

Delegating authority to environmental ministries does not always resolve conflicts satisfactorily, so governments are experimenting with various governance approaches to complement managerial ones. These range from ballots and polls, where engagement is passive and participants interact minimally, to adversarial processes that allow parties to redress grievances through formal legal procedures. They also include various experiments with intense interaction and deliberation aimed at negotiating decisions or allowing parties in potential conflict to provide structured input to them through participatory processes (91–95).

Inducing rule compliance. Effective governance requires that the rules of resource use are generally followed, with reasonable standards for tolerating modest violations. It is generally most effective to impose modest sanctions on first offenders, and gradually increase the severity of sanctions for those who do not learn from their first or second encounter (40, 96). Community-based institutions often use informal strategies for achieving compliance that rely on participants' commitment to rules and subtle social sanctions. Whether enforcement mechanisms are formal or informal, those who impose them must be seen as effective and legitimate by resource users or resistance and evasion will overwhelm the commons governance strategy.

Much environmental regulation in complex societies has been "command and control." Governments require or prohibit specific actions or technologies, with fines or jail terms possible to punish rule breakers. If sufficient resources are made available for monitoring and enforcement, such approaches are effective. But when governments lack the will or resources to protect "protected areas" such as parks (97–99), when major environmental damage comes from hard-to-detect "**nonpoint sources**," and when the need is to encourage innovation in behaviors or technologies rather than to require or prohibit familiar ones, command and control approaches are less effective. They are also economically inefficient in many circumstances (100–102).

Financial instruments can provide incentives to achieve compliance with environmental rules.

KEY TERM

Runoff that enters the air or the surface water, groundwater, and/or the oceans from widespread activities comes from **nonpoint sources** of pollution. In contrast, point source pollution has a specific source that can be easily identified. For example, a manufacturer that dumps waste directly into a creek is a point source. When pesticides used on many farms run into the groundwater, that is nonpoint source pollution. Sources of nonpoint pollution include agriculture and livestock, urban runoff, automobiles, land clearing, sewage, air pollution, and industrial waste. Of course, the difference is a matter of degree; with detailed examination, specific activities at specific locations can be identified as the origins of what is called nonpoint source pollution.

In recent years, market-based systems of tradable environmental allowances (TEAs) that define a limit to environmental withdrawals or emissions and permit free trade of allocated allowances under those limits have become popular (78, 101, 103). TEAs are one of the bases for the Kyoto agreement on climate change.

Economic theory and experience in some settings suggest that these mechanisms have substantial advantages over command and control (104–107). TEAs have exhibited good environmental performance and economic efficiency in the U.S. Sulfur Dioxide Allowance Market intended to reduce the prevalence of acid rain (108, 109) and the Lead Phasedown Program aimed at reducing the level of lead emissions (110). Crucial variables that differentiate these highly successful programs from less successful ones, such as chlorofluorocarbon production quota trading and the early EPA emission trading programs, include (i) the level of predictability of the stocks and flows, (ii) the number of users or producers who are regulated, (iii) the heterogeneity of the regulated users, and (iv) clearly defined and fully exchangeable permits (111).

TEAs, like all institutional arrangements, have notable limitations. TEA regimes tend to leave unprotected those resources not specifically covered by trading rules. For example, fish species caught as bycatch are often not covered (112). These regimes also tend to suffer when monitoring is difficult. For example, under the Kyoto Protocol, the question of whether geologically sequestered carbon will remain sequestered is difficult to answer. Problems can also occur with the initial allocation of allowances, especially when historic users, who may be called on to change their behavior most, have disproportionate power over allocation decisions (78, 113). TEAs and community-based systems appear to have opposite strengths and weaknesses (113), suggesting that institutions that combine aspects of both systems may work better than either approach alone. For example, the fisheries tradable permit system in New Zealand has added comanagement institutions to complement the market institutions (103, 114).

Voluntary approaches and those based on information disclosure have only begun to receive careful scientific attention as supplements to other tools (64, 79, 115–118). Success appears to depend on the existence of incentives that benefit leaders in volunteering over laggards and on the simultaneous use of other strategies, particularly ones that create incentives for compliance (79, 118–120). Difficulties of sanctioning pose major problems for international agreements (121–123).

Providing infrastructure. The importance of physical and technological infrastructure is often ignored. Infrastructure, including technology, determines the degree to which a commons can be exploited. The extent and quality of water systems determine how they distribute water, for example, and fishing technology has a decisive influence on the size of the catch. Infrastructure also determines the extent to which waste can be reduced in resource use, and the degree to which resource conditions and the behavior of human users can be effectively monitored. Indeed, the ability to choose institutional arrangements depends in part on infrastructure. In the absence of barbed-wire fences, for example, enforcing private property rights on grazing lands is expensive, but with barbed wire fences, it is relatively cheap (124).

Effective communication and transportation technologies are also of immense importance. Fishers who observe an unauthorized boat or harvesting technology can use a radio or cellular phone to alert others to illegal actions (125). Infrastructure also affects the links between local commons and regional and global systems. Good roads can provide food in bad times but can also open local resources to global markets, creating demand for resources that cannot be used locally (126). Institutional infrastructure is also important, including research, social capital, and multilevel rules, to coordinate between local and broader levels of governance (49, 127, 128).

Be prepared for change. Institutions must be designed to allow for adaptation because some current understanding is likely to be wrong, the required level of organization can shift, and bio-physical and social systems change. Fixed rules are likely to fail because they place too much confidence in the current state of knowledge, whereas systems that guard against the low-probability, high-consequence possibilities and allow for change may be suboptimal in the short run but prove wiser in the long run. This is a principal lesson of adaptive governance research (29, 32, 129).

An Illustration of the Challenge of Inducing Rule Compliance

Meeting these requirements is always a challenge. We illustrate by focusing on the problem of inducing rule compliance and comparing the experience of four national parks, three different biological communities, and a buffer zone contained within a single, large, and very famous biosphere reserve—the Maya Biosphere Reserve (MBR) in Guatemala (130). MBR (Figure 23) was created in 1990 by government decree to protect the remaining areas of pristine ecosystems in northern Guatemala (131). The region saw a marked advance of the agricultural frontier in the 1980s resulting from an aggressive policy of the central government to provide land to farmers from the south (132). MBR occupies over 21,000 km², equivalent to 19 percent of the Guatemalan territory, and represents the second-largest tract of tropical forest in the Western Hemisphere, after the Amazon (133). Much of the territory within the reserve has been seriously deforested and converted to agriculture and other uses. Figure 23 shows examples of several governance strategies and outcomes within MBR (134), but we concentrate here on two protected areas that have interesting institutional differences.

Tikal National Park is one of few protected areas in Guatemala to receive the full support of the government. The revenue from entry fees paid by tourists covers the entire budget for the park plus a surplus that goes to the Ministry of Culture and Sports. Directors of the park are held accountable by high-level officials for the successful protection of this source of government revenue. The park has permanent administrative and support staff, paid guards, and local residents hired to prevent forest fires. Although Tikal National Park is in better shape than many other parks (135), it faces multiple threats, especially from bordering communities in the form of forest fires ignited to transform the land for agricultural and livestock purposes and illegal extraction of forest products (136). The dark gray color of the park in Figure 23 shows the areas of stable forest.

Laguna del Tigre National Park and Biotope are managed by two different conservation agencies and include the largest protected wetland in Central America. The principal threats are human settlement and immigration, encroaching agriculture and livestock, oil prospecting and drilling, construction of roads and other infrastructures, and lawlessness (e.g., intentional setting of forest fires and drug trafficking and plantations). Like Tikal, Laguna del Tigre has been designated for the highest possible level of government protection. However, land speculation inside and outside the park and biotope, fueled by cattle ranchers, corrupt politicians, and other officials, has pushed illegal settlers deeper into the reserve, where they clear tree cover to establish new agricultural plots and homesteads. Numerous light gray patches within the park and biotope in Figure 23 reveal forest clearing (137). Oversight in Laguna del Tigre has been weak. The small and underpaid group of park rangers is unable to enforce the mandates assigned to them to protect the park from human settlements, illegal harvesting, and forest fires, and to sanction those who do not comply. It has not been unusual for people accused of violating conservation laws to threaten park officials to the point where the latter are afraid to enforce the law.

The Guatemalan cases illustrate that legally protecting threatened areas does not ensure rule compliance, especially when noncompliance is easy or profitable. Further illustrations of the

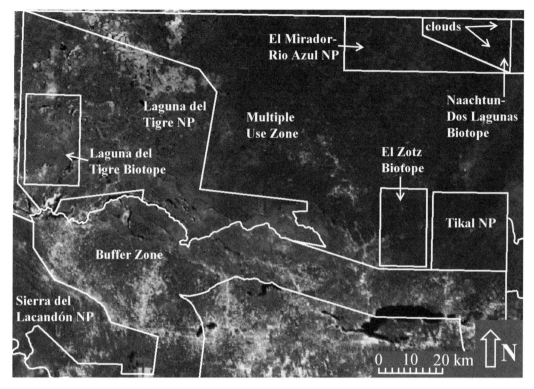

FIGURE 23.   This figure shows land-cover change and the numerous zones of the Maya Biosphere Reserve in northern Guatemala. The composite shows a uniform, dark gray color within Tikal, indicative of stable forest cover. El Mirador–Rio Azul National Park and Naachtun–Dos Laguna Biotope are also stable, due to inaccessibility. The other four potected areas have experienced extensive inroads of deforestation shown in light gray and white. Official designation as a protected area is not sufficient unless substantial investments are made in maintaining and enforcing boundaries. [Composite constructed by Glen Green, Edwin Castellanos, and Victor Hugo Ramos.

roles of institutions in forest protection are discussed in *Science* supplemental online materials (*68*) and in a new book (*138*).

Strategies for Meeting the Requirements of Adaptive Governance

The general principles for robust governance institutions for localized resources (Figure 24) are well established as a result of multiple empirical studies (*13, 40, 139–148*). Many of these also appear to be applicable to regional and global resources (*149*), although they are less well tested at those levels. Three of them seem to be particularly relevant for problems at broader scales.

Analytic deliberation. Well-structured dialogue involving scientists, resource users, and interested publics, and informed by analysis of key information about environmental and human-environment systems, appears critical. Such analytic deliberation (*76, 150–152*) provides improved information and the trust in it that is essential for information to be used effectively, builds social capital, and can allow for change and deal with inevitable conflicts well enough to produce consensus on governance rules. The negotiated 1994 U.S. regulation on disinfectant by-products in water that reached an interim consensus, including a decision to collect new information and reconsider the rule on that basis (*76*), is an excellent example of this approach.

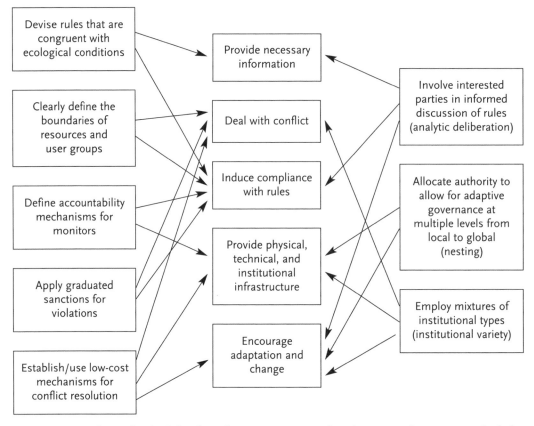

FIGURE 24. General principles for robust governance of environmental resources (shaded, left and right columns) and the governance requirements they help meet (center column) (13, 170). Each principle is relevant for meeting several requirements. Arrows indicate some of the most likely connections between principles and requirements. Principles in the right column may be particularly relevant for global and regional problems.

Nesting. Institutional arrangements must be complex, redundant, and nested in many layers (33, 153, 154). Simple strategies for governing the world's resources that rely exclusively on imposed markets or one-level, centralized command and control and that eliminate apparent redundancies in the name of efficiency have been tried and have failed. Catastrophic failures often have resulted when central governments have exerted sole authority over resources. Examples include the massive environmental degradation and impoverishment of local people in Indonesian Borneo (97); the increased rate of loss and fragmentation of high-quality habitat that occurred after creating the Wolong Nature Reserve in China (155); and the closing of the northern cod fishery along the eastern coast of Canada, which is partly attributable to the excessive quotas granted by the Canadian government (72). Governance should employ mixtures of institutional types (e.g., government bureaus as well as markets and community self-governance) that employ a variety of decision rules (about when and what resources should be harvested by whom) to change incentives, increase information, monitor use, and induce compliance (6, 64, 119). Innovative rule evaders can have more trouble with a multiplicity of rules than with a single type of rule.

Conclusion

Is it possible to govern such critical commons as the oceans and the climate? We remain guardedly optimistic. Thirty-five years ago it seemed that the "tragedy of the commons" was inevitable everywhere not owned privately or by a government. Systematic multidisciplinary research has, however, shown that a wide diversity of adaptive governance systems have been effective stewards of many resources. Sustained research coupled with an explicit view of national and international policies as experiments can yield the scientific knowledge necessary to design appropriate adaptive institutions.

Sound science is necessary for commons governance, but not sufficient. Too many strategies for governance of local commons are designed in capital cities or by donor agencies in ignorance of the state of the science and local conditions. The results are often tragic, but at least these tragedies are local. As the human footprint on Earth enlarges (156), humanity is challenged to develop and deploy understanding of broad-scale commons governance quickly enough to avoid the broad-scale tragedies that will otherwise ensue.

References and Notes

1. G. Hardin, *Science* 162, 1243 (1968).
2. See (6, 157). It was the paper most frequently cited as having the greatest career impact in a recent survey of biologists (158). A search performed by L. Wisen on 22 and 23 October 2003 on the Workshop Library Common-Pool Resources database (159) revealed that, before Hardin's paper, only 19 articles had been written in English-language academic literature with a specific reference to "commons," "common-pool resources," or "common property" in the title. Since then, attention to the commons has grown rapidly. Since 1968, a total of over 2,300 articles in that database contain a specific reference to one of these three terms in the title.
3. B. J. McCay, J. M. Acheson, *The Question of the Commons: The Culture and Ecology of Communal Resources* (Univ. of Arizona Press, Tucson, 1987).
4. P. Dasgupta, *Proc. Br. Acad.* 90, 165 (1996).
5. D. Feeny, F. Berkes, B. McCay, J. Acheson, *Hum. Ecol.* 18, 1 (1990).
6. Committee on the Human Dimensions of Global Change, National Research Council, *The Drama of the Commons*, E. Ostrom *et al.*, Eds. (National Academy Press, Washington, DC, 2002).
7. J. Platt, *Am. Psychol.* 28, 642 (1973).
8. J. G. Cross, M. J. Guyer, *Social Traps* (Univ. of Michigan Press, Ann Arbor, 1980).
9. R. Costanza, *Bioscience* 37, 407 (1987).
10. R. McC. Netting, *Balancing on an Alp: Ecological Change and Continuity in a Swiss Mountain Community* (Cambridge Univ. Press, Cambridge, 1981).
11. National Research Council, *Proceedings of the Conference on Common Property Resource Management* (National Academy Press, Washington, DC, 1986).
12. J.-M. Baland, J.-P. Platteau, *Halting Degradation of Natural Resources: Is There a Role for Rural Communities?* (Clarendon Press, Oxford, 1996).
13. E. Ostrom, *Governing the Commons: The Evolution of Institutions for Collective Action* (Cambridge Univ. Press, New York, 1990).
14. E. Ostrom, *Understanding Institutional Diversity* (Princeton University Press, Princeton, NJ, 2005).
15. R. A. Myers, B. Worm, *Nature* 423, 280 (2003).
16. A. C. Finlayson, *Fishing for Truth: A Sociological Analysis of Northern Cod Stock Assessments from 1987 to 1990* (Institute of Social and Economic Research, Memorial Univ. of Newfoundland, St. Johns, Newfoundland, 1994).
17. S. Hanna, in *Northern Waters: Management Issues and Practice*, D. Symes, Ed. (Blackwell, London, 1998), pp. 25–35.
18. J. Acheson, *Capturing the Commons: Devising Institutions to Manage the Maine Lobster Industry* (Univ. Press of New England, Hanover, NH, 2003).
19. J. A. Wilson, P. Kleban, J. Acheson, M. Metcalfe, *Mar. Policy* 18, 291 (1994).
20. J. Wilson, personal communication.
21. S. Weiner, J. Maxwell, in *Dimensions of Managing Chlorine in the Environment*, report of the MIT/Norwegian Chlorine Policy Study (MIT, Cambridge, MA, 1993).
22. U. Weber, *UNESCO Courier*, June 2000, p. 9.

23. M. Verweij, *Transboundary Environmental Problems and Cultural Theory: The Protection of the Rhine and the Great Lakes* (Palgrave, New York, 2000).

24. C. Dieperink, *Water Int.* 25, 347 (2000).

25. E. Parson, *Protecting the Ozone Layer: Science and Strategy* (Oxford Univ. Press, New York, 2003).

26. E. Ostrom, C. D. Becker, *Annu. Rev. Ecol. Syst.* 26, 113 (1995).

27. P. C. Stern, *Science* 260, 1897 (1993).

28. E. Ostrom, J. Burger, C. B. Field, R. B. Norgaard, D. Policansky, *Science* 284, 278 (1999).

29. We refer to *adaptive governance* rather than *adaptive management* (*32, 129*) because the idea of governance conveys the difficulty of control, the need to proceed in the face of substantial uncertainty, and the importance of dealing with diversity and reconciling conflict among people and groups who differ in values, interests, perspectives, power, and the kinds of information they bring to situations (*150, 160–163*). Effective environmental governance requires an understanding of both environmental systems and human-environment interactions (*27, 84, 164, 165*).

30. E. Schlager, W. Blomquist, S. Y. Tang, *Land Econ.* 70, 294 (1994).

31. J. H. Brander, M. S. Taylor, *Am. Econ. Rev.* 88, 119 (1998).

32. L. H. Gunderson, C. S. Holling, *Panarchy: Understanding Transformations in Human and Natural Systems* (Island Press, Washington, DC, 2001).

33. M. Janssen, *Complexity and Ecosystem Management* (Elgar, Cheltenham, UK, 2002).

34. R. Putnam, *Bowling Alone: The Collapse and Revival of American Community* (Simon and Schuster, New York, 2001).

35. A. Bebbington, *Geogr. J.* 163, 189 (1997).

36. R. Frank, *Passions Within Reason: The Strategic Role of the Emotions* (Norton, New York, 1988).

37. J. Pretty, *Science* 302, 1912 (2003).

38. J. Burger, E. Ostrom, R. B. Norgaard, D. Policansky, B. D. Goldstein, Eds., *Protecting the Commons: A Framework for Resource Management in the Americas* (Island Press, Washington, DC, 2001).

39. C. Gibson, J. Williams, E. Ostrom, in preparation. *World Dev.* 33, 273 (2005).

40. M. S. Weinstein, *Georgetown Int. Environ. Law Rev.* 12, 375 (2000).

41. R. Meinzen-Dick, K. V. Raju, A. Gulati, *World Dev.* 30, 649 (2002).

42. E. L. Miles *et al.*, Eds., *Environmental Regime Effectiveness: Confronting Theory with Evidence* (MIT Press, Cambridge, MA, 2001).

43. C. Gibson, M. McKean, E. Ostrom, Eds., *People and Forests* (MIT Press, Cambridge, MA, 2000).

44. S. Krech III, *The Ecological Indian: Myth and History* (Norton, New York, 1999).

45. For relevant bibliographies, see (*159, 166*).

46. D. C. North, *Structure and Change in Economic History* (North, New York, 1981).

47. R. Robertson, *Globalization: Social Theory and Global Culture* (Sage, London, 1992).

48. O. R. Young, Ed., *The Effectiveness of International Environmental Regimes* (MIT Press, Cambridge, MA, 1999).

49. O. R. Young, *The Institutional Dimensions of Environmental Change: Fit, Interplay, and Scale* (MIT Press, Cambridge, MA, 2002).

50. R. Keohane, E. Ostrom, Eds., *Local Commons and Global Interdependence* (Sage, London, 1995).

51. J. S. Lansing, *Priests and Programmers: Technologies of Power in the Engineered Landscape of Bali* (Princeton Univ. Press, Princeton, NJ, 1991).

52. J. Wunsch, D. Olowu, Eds., *The Failure of the Centralized State* (Institute for Contemporary Studies Press, San Francisco, CA, 1995).

53. N. Dolöak, E. Ostrom, Eds., *The Commons in the New Millennium: Challenges and Adaptation* (MIT Press, Cambridge, MA, 2003).

54. Association of American Geographers Global Change and Local Places Research Group, *Global Change and Local Places: Estimating, Understanding, and Reducing Greenhouse Gases* (Cambridge Univ. Press, Cambridge, 2003).

55. S. Karlsson, thesis, Linköping University, Sweden (2000).

56. R. Keohane, M. A. Levy, Eds., *Institutions for Environmental Aid* (MIT Press, Cambridge, MA, 1996).

57. O. S. Stokke, *Governing High Seas Fisheries: The Interplay of Global and Regional Regimes* (Oxford Univ. Press, London, 2001).

58. A. Underdal, K. Hanf, Eds., *International Environmental Agreements and Domestic Politics: The Case of Acid Rain* (Ashgate, Aldershot, England, 1998).

59. W. Clark, R. Munn, Eds., *Sustainable Development of the Biosphere* (Cambridge Univ. Press, New York, 1986).

60. B. L. Turner II *et al.*, *Global Environ. Change* 1, 14 (1991).

61. T. Dietz, T. R. Burns, *Acta Sociol.* 35, 187 (1992).

62. T. Dietz, E. A. Rosa, in *Handbook of Environmental Sociology*, R. E. Dunlap, W. Michelson, Eds. (Greenwood Press, Westport, CT, 2002), pp. 370–406.

63. A. P. Vayda, in *Ecology in Practice*, F. di Castri *et al.*, Eds. (Tycooly, Dublin, 1984).

64. Committee on the Human Dimensions of Global Change, National Research Council, *New Tools for Environmental Protection: Education, Information, and Voluntary Measures*, T. Dietz, P. C. Stern, Eds. (National Academy Press, Washington, DC, 2002).

65. M. Auer, *Policy Sci.* 33, 155 (2000).

66. D. H. Cole, *Pollution and Property: Comparing Ownership Institutions for Environmental Protection* (Cambridge Univ. Press, Cambridge, 2002).

67. F. Berkes, J. Colding, C. Folke, Eds., *Navigating Social-Ecological Systems: Building Resilience for Complexity and Change* (Cambridge Univ. Press, Cambridge, 2003).

68. Supporting Online Material in T. Dietz, E. Ostrom, P. C. Stern, *Science* 302, 1907 (2003); available at www.sciencemag.org/cgi/content/full/302/5652/1907/DC1.

69. K. J. Willis, R. J. Whittaker, *Science* 295, 1245 (2002).

70. Kirby Task Force on Atlantic Fisheries, *Navigating Troubled Waters: A New Policy for the Atlantic Fisheries* (Department of Fisheries and Oceans, Ottawa, 1982).

71. G. Barrett, A. Davis, *J. Can. Stud.* 19, 125 (1984).

72. A. C. Finlayson, B. McCay, in *Linking Social and Ecological Systems*, F. Berkes, C. Folke, Eds. (Cambridge Univ. Press, Cambridge, 1998), pp. 311–338.

73. J. A. Wilson, R. Townsend, P. Kleban, S. McKay, J. French, *Ocean Shoreline Manage.* 13, 179 (1990).

74. C. Martin, *Fisheries* 20, 6 (1995).

75. Committee on Risk Perception and Communication, National Research Council, *Improving Risk Communication* (National Academy Press, Washington, DC, 1989).

76. Committee on Risk Characterization and Commission on Behavioral and Social Sciences and Education, National Research Council, *Understanding Risk: Informing Decisions in a Democratic Society*, P. C. Stern, H. V. Fineberg, Eds. (National Academy Press, Washington, DC, 1996).

77. Panel on Human Dimensions of Seasonal-to-Interannual Climate Variability, Committee on the Human Dimensions of Global Change, National Research Council, *Making Climate Forecasts Matter*, P. C. Stern, W. E. Easterling, Eds. (National Academy Press, Washington, DC, 1999).

78. T. Tietenberg, in *The Drama of the Commons*, Committee on the Human Dimensions of Global Change, National Research Council, E. Ostrom *et al.*, Eds. (National Academy Press, Washington, DC, 2002), pp. 233–257.

79. T. Tietenberg, D. Wheeler, in *Frontiers of Environmental Economics*, H. Folmer, H. Landis Gabel, S. Gerking, A. Rose, Eds. (Elgar, Cheltenham, UK, 2001), pp. 85–120.

80. J. Thøgerson, in *New Tools for Environmental Protection: Education, Information, and Voluntary Measures*, T. Dietz, P. C. Stern, Eds. (National Academy Press, Washington, DC, 2002), pp. 83–104.

81. J. A. Wilson, in *The Drama of the Commons*, Committee on the Human Dimensions of Global Change, National Research Council, E. Ostrom *et al.*, Eds. (National Academy Press, Washington, DC, 2002), pp. 327–360.

82. R. Moss, S. H. Schneider, in *Guidance Papers on the Cross-Cutting Issues of the Third Assessment Report of the IPCC*, R. Pachauri, T. Taniguchi, K. Tanaka, Eds.

(World Meteorological Organization, Geneva, Switzerland, 2000), pp. 33–51.

83. B. J. McCay, in *The Drama of the Commons*, Committee on the Human Dimensions of Global Change, National Research Council, E. Ostrom *et al.*, Eds. (National Academy Press, Washington, DC, 2002), pp. 361–402.

84. Board on Sustainable Development, National Research Council, *Our Common Journey: A Transition Toward Sustainability* (National Academy Press, Washington, DC, 1999).

85. Committee on Noneconomic and Economic Value of Biodiversity, National Research Council, *Perspectives on Biodiversity: Valuing Its Role in an Everchanging World* (National Academy Press, Washington, DC, 1999).

86. W. M. Adams, D. Brockington, J. Dyson, B. Vira, *Science* 302, 1915 (2003).

87. P. C. Stern, *Policy Sci.* 24, 99 (1991).

88. V. Ostrom, *Public Choice* 77, 163 (1993).

89. R. Costanza, M. Ruth, in *Institutions, Ecosystems, and Sustainability*, R. Costanza, B. S. Low, E. Ostrom, J. Wilson, Eds. (Lewis Publishers, Boca Raton, FL, 2001), pp. 169–178.

90. F. H. Sklar, M. L. White, R. Costanza, *The Coastal Ecological Landscape Spatial Simulation (CELSS) Model* (U.S. Fish and Wildlife Service, Washington, DC, 1989).

91. O. Renn, T. Webler, P. Wiedemann, Eds., *Fairness and Competence in Citizen Participation: Evaluating Models for Environmental Discourse* (Kluwer Academic Publishers, Dordrecht, Netherlands, 1995).

92. R. Gregory, T. McDaniels, D. Fields, *J. Policy Anal. Manage.* 20, 415 (2001).

93. T. C. Beierle, J. Cayford, *Democracy in Practice: Public Participation in Environmental Decisions* (Resources for the Future, Washington, DC, 2002).

94. W. Leach, N. Pelkey, P. Sabatier, *J. Policy Anal. Manage.* 21, 645 (2002).

95. R. O'Leary, L. B. Bingham, Eds., *The Promise and Performance of Environmental Conflict Resolution* (Resources for the Future, Washington, DC, 2003).

96. E. Ostrom, R. Gardner, J. Walker, Eds., *Rules, Games, and Common-Pool Resources* (Univ. of Michigan Press, Ann Arbor, 1994).

97. L. M. Curran, S. N. Trigg, A. K. McDonald, D. Astiani, Y. M. Hardiono, P. Siregar, I. Caniago, and E. Kasischke, *Science*, 303, 1000 (2004).

98. J. Liu *et al.*, *Science* 300, 1240 (2003).

99. R. W. Sussman, G. M. Green, L. K. Sussman, *Hum. Ecol.* 22, 333 (1994).

100. F. Berkes, C. Folke, Eds., *Linking Social and Ecological Systems: Management Practices and Social Mechanisms* (Cambridge Univ. Press, Cambridge, 1998).

101. G. M. Heal, *Valuing the Future: Economic Theory and Sustainability* (Colombia Univ. Press, New York, 1998).

102. B. G. Colby, in *The Handbook of Environmental Economics*, D. Bromley, Ed. (Blackwell Publishers, Oxford, 1995), pp. 475–502.

103. T. Yandle, C. M. Dewees, in *The Commons in the New Millennium: Challenges and Adaptation*, N. Dolöak, E. Ostrom, Eds. (MIT Press, Cambridge, MA, 2003), pp. 101–128.

104. G. Libecap, *Contracting for Property Rights* (Cambridge Univ. Press, Cambridge, 1990).

105. R. D. Lile, D. R. Bohi, D. Burtraw, *An Assessment of the EPA's SO$_2$ Emission Allowance Tracking System* (Resources for the Future, Washington, DC, 1996).

106. R. N. Stavins, *J. Econ. Perspect.* 12, 133 (1998).

107. J. E. Wilen, *J. Environ. Econ. Manage.* 39, 309 (2000).

108. A. D. Ellerman, R. Schmalensee, P. L. Joskow, J. P. Montero, E. M. Bailey, *Emissions Trading Under the U.S. Acid Rain Program* (MIT Center for Energy and Environmental Policy Research, Cambridge, MA, 1997).

109. E. M. Bailey, "Allowance trading activity and state regulatory rulings" (Working Paper 98-005, MIT Emissions Trading, Cambridge, MA, 1998).

110. B. D. Nussbaum, in *Climate Change: Designing a Tradeable Permit System* (OECD, Paris, 1992), pp. 22–34.

111. N. Dolšak, thesis, Indiana University, Bloomington, IN (2000).

112. S. L. Hsu, J. E. Wilen, *Ecol. Law Q.* 24, 799 (1997).

113. C. Rose, in *The Drama of the Commons*, Committee on the Human Dimensions of Global Change, National Research Council, E. Ostrom *et al.*, Eds. (National Academy Press, Washington, DC, 2002), pp. 233–257.

114. E. Pinkerton, *Co-operative Management of Local Fisheries* (Univ. of British Columbia Press, Vancouver, 1989).

115. A. Prakash, *Bus. Strategy Environ.* 10, 286 (2001).

116. J. Nash, in *New Tools for Environmental Protection: Education, Information and Voluntary Measures*, T. Dietz, P. C. Stern, Eds. (National Academy Press, Washington, DC, 2002), pp. 235–252.

117. J. A. Aragón-Correa, S. Sharma, *Acad. Manage. Rev.* 28, 71 (2003).

118. A. Randall, in *New Tools for Environmental Protection: Education, Information and Voluntary Measures*, T. Dietz, P. C. Stern, Eds. (National Academy Press, Washington, DC, 2002), pp. 311–318.

119. G. T. Gardner, P. C. Stern, *Environmental Problems and Human Behavior* (Allyn and Bacon, Needham Heights, MA, 1996).

120. P. C. Stern, *J. Consum. Policy* 22, 461 (1999).

121. S. Hanna, C. Folke, K.-G. Mäler, *Rights to Nature* (Island Press, Washington, DC, 1996).

122. E. Weiss, H. Jacobson, Eds., *Engaging Countries: Strengthening Compliance with International Environmental Agreements* (MIT Press, Cambridge, MA, 1998).

123. A. Underdal, *The Politics of International Environmental Management* (Kluwer Academic Publishers, Dordrecht, Netherlands, 1998).

124. A. Krell, *The Devil's Rope: A Cultural History of Barbed Wire* (Reaktion, London, 2002).

125. S. Singleton, *Constructing Cooperation: The Evolution of Institutions of Comanagement* (Univ. of Michigan Press, Ann Arbor, 1998).

126. E. Moran, Ed., *The Ecosystem Approach in Anthropology: From Concept to Practice* (Univ. of Michigan Press, Ann Arbor, 1990).

127. M. Janssen, J. M. Anderies, E. Ostrom, paper presented at the Workshop on Resiliency and Change in Ecological Systems, Santa Fe Institute, Santa Fe, NM, 25 to 27 October 2003.

128. T. Princen, *Global Environ. Polit.* 3, 33 (2003).

129. K. Lee, *Compass and Gyroscope* (Island Press, Washington, DC, 1993).

130. Text in this illustration was adapted from text drafted by Lilian Marquez-Barrientos and Edwin Castellanos (*68*).

131. G. G. Stuart, *Nat. Geogr. Mag.*, 182, 94 (1992).

132. E. G. Katz, *Land Econ.*, 76, 114 (2000).

133. S. Elias, *Petén y los retos para el desarrollo sostenible in encuentro internacional de investigadores: nuevas perspectivas de desarrollo sostenible en Petén* (Facultad Latinoamericana de Ciencias Sociales, Guatemala City, Guatemala, 2000).

134. Figure 23 was produced and interpreted by a group of scholars associated with the Center for the Study of Institutions, Population, and Environmental Change at Indiana University.

135. S. A. Sader, D. J. Hayes, J. A. Hepinstal, M. Coan, C. Soza, *Int. J Remote Sensing*, 22, 1937 (2001).

136. ParksWatch, Tikal National Park, www.parks watch.org/parkprofile.php?l=eng&country=gua&park=tinp&page=con.

137. A recent report in *U.S. News & World Report* (*171*) strongly verifies the satellite data and illuminates the causes of continuing deforestation in the Laguna del Tigre National Park since the most recent satellite image in Figure 23, taken in 2000. The reporter wrote: "Since the end of the civil war in 1996 desperately poor farmers and rich cattle ranchers have been pouring into this vast, virgin rain forest. In the past five years, more than 5,000 homesteaders have illegally built homes and set fires to clear the land for corn and cattle. In 2003, [Dr. David] Freidel's first year of excavation here [at the "El Peru" archaeological site], the flames got so close—about 2 miles away—that he had to pull workers off the ruins to dig fire lines to save the camp. . . . The ground fires have so far burned an estimated 40 percent of the park, which is supposed to protect a lake that is one of the most important wetlands in the world" (online at www.usnews.com/usnews/culture/articles/050627/27profile_3.htm).

138. E. F. Moran, E. Ostrom, Eds., *Seeing the Forest and the Trees: Human-Environment Interactions in Forest Ecosystems* (MIT Press, Cambridge, MA, 2005).

139. C. L. Abernathy, H. Sally, *J. Appl. Irrig. Stud.* 35, 177 (2000).

140. A. Agrawal, in *The Drama of the Commons*, Committee on the Human Dimensions of Global Change, National Research Council, E. Ostrom *et al.*, Eds. (National Academy Press, Washington, DC, 2002), pp. 41–85.

141. P. Coop, D. Brunckhorst, *Aust. J. Environ. Manage.* 6, 48 (1999).

142. D. S. Crook, A. M. Jones, *Mt. Res. Dev.* 19, 79 (1999).

143. D. J. Merrey, in *Irrigation Management Transfer*, S. H. Johnson, D. L. Vermillion, J. A. Sagardoy, Eds. (International Irrigation Management Institute, Colombo, Sri Lanka and the Food and Agriculture Organisation, Rome, 1995).

144. C. E. Morrow, R. W. Hull, *World Dev.* 24, 1641 (1996).

145. T. Nilsson, thesis, Royal Institute of Technology, Stockholm, Sweden (2001).

146. N. Polman, L. Slangen, in *Environmental Co-operation and Institutional Change*, K. Hagedorn, Ed. (Elgar, Northampton, MA, 2002).

147. A. Sarker, T. Itoh, *Agric. Water Manage.* 48 (no. 8), 9 (2001).

148. C. Tucker, *Praxis* 15, 47 (1999).

149. R. Costanza *et al.*, *Science* 281, 198 (1998).

150. T. Dietz, P. C. Stern, *Bioscience* 48, 441 (1998).

151. E. Rosa, A. M. McWright, O. Renn, "The risk society: Theoretical frames and state management challenges" (Dept. of Sociology, Washington State Univ., Pullman, WA, 2003).

152. National Research Council, Panel on Social and Behavioral Science Research Priorities for Environmental Decision Making, *Decision Making for the Environment: Social and Behavioral Science Research Priorities*, G. B. Brewer, P.C. Stern, Eds. (National Academy Press, Washington DC, 2005).

153. S. Levin, *Fragile Dominion: Complexity and the Commons* (Perseus Books, Reading, MA, 1999).

154. B. Low, E. Ostrom, C. Simon, J. Wilson, in *Navigating Social-Ecological Systems: Building Resilience for Complexity and Change*, F. Berkes, J. Colding, C. Folke, Eds. (Cambridge Univ. Press, New York, 2003), pp. 83–114.

155. J. Liu *et al.*, *Science* 292, 98 (2001).

156. R. York, E. A. Rosa, T. Dietz, *Am. Sociol. Rev.* 68, 279 (2003).

157. G. Hardin, *Science* 280, 682 (1998).

158. G. W. Barrett, K. E. Mabry, *Bioscience* 52 (no. 28), 2 (2002).

159. C. Hess, *The Comprehensive Bibliography of the Commons*, database available online at www.indiana.edu/_iascp/Iforms/searchcpr.html.

160. V. Ostrom, *The Meaning of Democracy and the Vulnerability of Democracies* (Univ. of Michigan Press, Ann Arbor, 1997).

161. M. McGinnis, Ed., *Polycentric Governance and Development: Readings from the Workshop in Political Theory and Policy Analysis* (Univ. of Michigan Press, Ann Arbor, 1999).

162. M. McGinnis, Ed., *Polycentric Games and Institutions: Readings from the Workshop in Political Theory and Policy Analysis* (Univ. of Michigan Press, Ann Arbor, 2000).

163. T. Dietz, *Hum. Ecol. Rev.* 10, 60 (2003).

164. R. Costanza, B. S. Low, E. Ostrom, J. Wilson, Eds., *Institutions, Ecosystems, and Sustainability* (Lewis Publishers, New York, 2001).

165. Committee on the Human Dimensions of Global Change, National Research Council, *Global Environmental Change: Understanding the Human Dimensions*, P. C. Stern, O. R. Young, D. Druckman, Eds. (National Academy Press, Washington, DC, 1992).

166. C. Hess, *A Comprehensive Bibliography of Common-Pool Resources* (CD-ROM, Workshop in Political Theory and Policy Analysis, Indiana Univ., Bloomington, 1999).

167. Groundfish data were compiled by D. Gilbert (Maine Department of Marine Resources) with data from the National Marine Fisheries Service. Lobster data were compiled by C. Wilson (Maine Department of Marine Resources). J. Wilson (University of Maine) worked with the authors in the preparation of this figure.

168. United Nations Environment Programme, *Production and Consumption of Ozone Depleting Substances, 1986–1998* (United Nations Environment Programme Ozone Secretariat, Nairobi, Kenya, 1999).

169. World Resources Institute, *World Resources 2002–2004: EarthTrends Data CD* (World Resources Institute, Washington, DC, 2003).

170. P. C. Stern, T. Dietz, E. Ostrom, *Environ. Pract.* 4, 61 (2002).

171. K. Clark, *U.S. News & World Report*, 27 June 2005, pp. 54–57 (available online at http://www.usnews.com/usnews/culture/articles/050627/27profile_3.htm.

172. We thank R. Andrews, G. Daily, J. Hoehn, K. Lee, S. Levin, G. Libecap, V. Ruttan, T. Tietenberg, J. Wilson, and O. Young for their comments on earlier drafts; and G. Laasby, P. Lezotte, C. Liang, and L. Wisen for providing assistance and J. Broderick for her extensive editing assistance. Supported in part by NSF grants BCS-9906253 and SBR-9521918, NASA grant NASW-01008, the Ford Foundation, and the MacArthur Foundation.

Social Capital
and the Collective
Management of Resources

JULES PRETTY

The proposition that natural resources need protection from the destructive actions of people is widely accepted. Yet communities have shown in the past, and increasingly show today, that they can collaborate for long-term resource management. The term **social capital** captures the idea that social bonds and norms are critical for sustainability. Where social capital is high in formalized groups, people have the confidence to invest in collective activities, knowing that others will do so too. Some 0.4 to 0.5 million groups have been established since the early 1990s for watershed, forest, irrigation, pest, wildlife, fishery, and microfinance management. These offer a route to sustainable management and governance of common resources.

From Malthus to Hardin and beyond, analysts and policymakers have widely come to accept that

natural resources need to be protected from the destructive, yet apparently rational, actions of people. The compelling logic is that people inevitably harm natural resources as they use them, and more people therefore do more harm. The likelihood of this damage being greater where natural resources are commonly owned is further increased by suspicions that people tend to free-ride, both by overusing and underinvesting in the maintenance of resources. As our global numbers have increased, and as incontrovertible evidence of harm to water, land, and atmospheric resources has emerged, so the choices seem to be starker. Either we regulate to prevent further harm—in Hardin's words (1), to engage in mutual coercion mutually agreed upon—or we press ahead with enclosure and privatization to increase the likelihood that resources will be more carefully managed.

These concepts have influenced many policymakers and practitioners. They have led, for example, to the popular wilderness myth (2)—

This article first appeared in *Science* (12 December 2003: Vol. 302, no. 5652). It has been revised and updated for this edition.

KEY TERMS

There are five types of **capital**.

Natural capital produces environmental goods and services, and comprises food (both farmed and harvested or caught from the wild), wood, and fiber; water supply and regulation; treatment, assimilation, and decomposition of wastes; nutrient cycling and fixation; soil formation; biological control of pests; climate regulation; wildlife habitats; storm protection and flood control; carbon sequestration; pollination; and recreation and leisure.

Social capital yields a flow of mutually beneficial collective action, contributing to the cohesiveness of people in their societies. The social assets composing social capital include norms, values, and attitudes that predispose people to cooperate; relations of trust, reciprocity, and obligations; and common rules and sanctions mutually agreed or handed down. These are connected and structured in networks and groups.

Human capital is the total capability residing in individuals, based on their stock of knowledge, skills, health, and nutrition. It is enhanced by access to services that provide these, such as schools, medical services, and adult training. People's productivity is increased by their capacity to interact with productive technologies and with other people. Leadership and organizational skills are particularly important in making other resources more valuable.

Physical capital is the store of human-made material resources, and comprises buildings, such as housing and factories, market infrastructure, irrigation works, roads and bridges, tools and tractors, communications, and energy and transportation systems, that make labor more productive.

Financial capital is more of an accounting concept, as it serves as a facilitating role rather than as a source of productivity in and of itself. It represents accumulated claims on goods and services, built up through financial systems that gather savings and issue credit, such as pensions, remittances, welfare payments, grants, and subsidies.

that many ecosystems are pristine and have emerged independent of the actions of local people, whether positive or negative. Empty, idle, and "natural" environments need protection from harmful large-scale developers, loggers, and

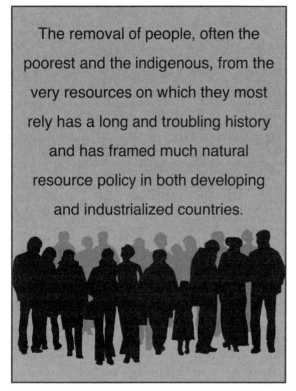

The removal of people, often the poorest and the indigenous, from the very resources on which they most rely has a long and troubling history and has framed much natural resource policy in both developing and industrialized countries.

ranchers, as well as from farmers, hunters, and gatherers (3). Since the first national park was set up at Yellowstone in 1872, some 12,750 protected areas of greater than 1,000 hectares have been established worldwide.

Of the 7,322 protected areas in developing countries where many people rely on wild resources for food, fuel, medicine, and feed, 30 percent covering 6 million square kilometers are strictly protected, permitting no use of resources. (4) The removal of people, often the poorest and the indigenous (5), from the very resources on which they most rely has a long and troubling history and has framed much natural resource policy in both developing and industrialized countries (6). Yet common property resources remain immensely valuable for many people, and exclusion can be costly for them. In India, for example, common resources have been

estimated to contribute some US$5 billion per year to the income of the rural poor (7).

An important question is, could local people play a positive role in conserving and managing resources? And if so, how best can unfettered private actions be mediated in favor of the common good? Though some communities have long been known to manage common resources such as forests and grazing lands effectively over long periods without external help (8), recent years have seen the emergence of local groups as an effective option instead of strict regulation or enclosure. This "third way" has been shaped by theoretical developments in the governance of the commons and in thinking on social capital (9, 10).

These groups are indicating that, given good knowledge about local resources; appropriate institutional, social, and economic conditions (11); and processes that encourage careful deliberation (12), communities can work together collectively to use natural resources sustainably over the long term (13).

Social Capital and Local Resource Management Groups

The term *social capital* captures the idea that social bonds and norms are important for people and communities (14). It emerged as a term after detailed analyses of the effects of social cohesion on regional incomes, civil society, and life expectancy (15–17). As social capital lowers the transaction costs of working together, it facilitates cooperation. People have the confidence to invest in collective activities, knowing that others will also do so. They are also less likely to engage in unfettered private actions with negative outcomes, such as resource degradation (18, 19). Four features are important: relations of trust; reciprocity and exchanges; common rules, norms, and sanctions; and connectedness in networks and groups.

Trust lubricates cooperation. It reduces the transaction costs between people and so liberates resources. Instead of having to invest in monitoring others, individuals are able to trust them to act as expected. This saves money and time. It can also create a social obligation—trusting someone engenders reciprocal trust. There are two types of trust: the trust we have in individuals whom we know and the trust we have in those we do not know, but which arises because of our confidence in a known social structure. Trust takes time to build but is easily broken, and when a society is pervaded by distrust, cooperative arrangements are unlikely to emerge (20).

Reciprocity and exchanges also increase trust. There are two types of reciprocity: *specific reciprocity*, which refers to simultaneous exchanges of items of roughly equal value; and *diffuse reciprocity*, which is a continuing relationship of exchange that at any given time may not be met but eventually is repaid and balanced (14, 15). This contributes to the development of long-term obligations between people.

Common rules, norms, and sanctions are the mutually agreed or handed-down norms of behavior that place group interests above those of individuals. These are sometimes called the rules of the game (21), and they give individuals the confidence to invest in collective or group activities, knowing that others will also do so. Individuals can take responsibility and ensure that their rights are not infringed. Mutually agreed sanctions ensure that those who break the rules know that they will be punished. Formal rules are those set out by authorities, such as laws and regulations, whereas informal ones are those individuals use to shape their own everyday behavior. Norms are, by contrast, preferences, and they indicate how individuals should act. A high social capital implies high "internal morality," with individuals balancing individual rights with collective responsibilities.

Connectedness, networks, and groups are a vital aspect of social capital. Three types of connectedness are important: bonding, bridging, and linking types of social capital (22). *Bonding* describes the links between people with similar outlooks and objectives and is manifested in different types of groups at the local level—from

guilds and mutual aid societies to sports clubs and credit groups, from forest or fishery management groups to literary societies and mothers' groups. *Bridging* describes the capacity of groups to make links with others that may have different views, particularly across communities. Such horizontal connections can sometimes lead to the establishment of new platforms and apex organizations that represent large numbers of individuals and groups. *Linking* describes the ability of groups to engage vertically with external agencies, either to influence their policies or to draw down on resources.

There is growing evidence that high social capital is associated with improved economic and social well-being. Households with greater connectedness have been shown to have higher incomes, such as in Tanzania, India, and China (23–25), better health (26), improved educational achievements (27), and better social cohesion and more constructive links with government (28).

Communities also do not always have the knowledge to appreciate that what they are doing may be harmful. For instance, it is common for fishing communities to believe that fish stocks are not being eroded, even though the scientific evidence indicates otherwise. Local groups may need the support of higher level authorities.

There is a danger, of course, of appearing too optimistic about local groups and their capacity to deliver economic and environmental benefits. It is important to be aware of the divisions and differences within and between communities,

TABLE 5. Social capital formation in selected agricultural and rural resource management sectors (since the early 1990s). This table suggests that 455,000 to 520,000 groups have been formed. Additional groups have been formed in farmers' research, fishery, and wildlife programs in a wide variety of countries (21).

	Countries and programs	Local groups (thousand)
Watershed and catchment groups	Australia (4500 Landcare groups containing about one-third of all farmers), Brazil (15,000 to 17,000 microbacias groups), Guatemala, and Honduras (700 to 1100 groups), India (30,000 groups in both state government and nongovernmental organization programs), Kenya (3000 to 4500 Ministry of Agriculture catchment committees), United States (1000 farmer-led watershed initiatives)	54 to 58
Irrigation water users' groups	Sri Lanka, Nepal, India, Philippines, and Pakistan (water users' groups as part of government irrigation programs)	58
Microfinance institutions	Bangladesh (Grameen Bank and Proshika), Nepal, India, Sri Lanka, Vietnam, China, Philippines, Fiji, Tonga, Solomon Islands, Papua New Guinea, Indonesia, and Malaysia	252 to 295
Joint and participatory forest management	India and Nepal (joint forest management and forest protection committees)	73
Integrated pest management	Indonesia, Vietnam, Bangladesh, Sri Lanka, China, Philippines, and India (farmers trained in farmer field schools)	18 to 36

and to realize how conflicts can result in environmental damage. Not all forms of social relations are necessarily good for everyone in a community. A society may be well organized, have strong institutions, and have embedded reciprocal mechanisms, but be based not on trust but on fear and power, such as in feudal, racist, and unjust societies (29). Formal rules and norms can also trap people within harmful social arrangements. Again, a system may appear to have high levels of social assets, with strong families and religious groups, but contain abused individuals or those in conditions of slavery or other exploitation. Some associations can also act as obstacles to the emergence of sustainability, encouraging conformity, perpetuating adversity and inequity, and allowing some individuals to get others to act in ways that suit only themselves.

Do these ideas about social capital work in practice? First, there is evidence that high social capital is associated with improved economic and social well-being. Households with greater connectedness tend to have higher incomes, better health, higher educational achievements, and more constructive links with government (4, 9,

15, 16, 30). What, then, can be done to develop appropriate forms of social organization that structurally suit natural resource management? Collective resource management programs that seek to build trust, develop new norms, and help form groups have become increasingly common, and such programs are variously described by the terms community-, participatory-, joint-, decentralized-, and co-management. They have been effective in several sectors, including watershed, forest, irrigation, pest, wildlife, fishery, farmers' research, and microfinance management (Table 5). Since the early 1990s, some 400,000 to 500,000 new local groups were established in varying environmental and social contexts (18), mostly evolving to be of similar small size, typically with 20 to 30 active members, putting total involvement at some 8 to 15 million households. The majority continue to be successful and show the inclusive characteristics identified as vital for improving community well-being (31), and evaluations have confirmed that there are positive ecological and economic outcomes, including for watersheds (30), forests (32), and pest management (33, 34).

KEY TERM

A **co-management regime** is the sharing of natural resource management between local representatives, their institutions, and the formal institutions at local level (e.g., provincial governments). Local participants can have varying degrees of decision-making power, from advisory to shared jurisdiction. Such arrangements may be termed **community, participatory, joint,** or **decentralized regimes**. They are most commonly used in forestry and fisheries management, particularly in developing countries, and in some remote and highly fishery-dependent regions of North America.

Further Challenges

The formation, persistence, and effects of new groups suggests that new configurations of social and human relations could be prerequisites for long-term improvements in natural resources. Regulations and economic incentives play an important role in encouraging changes in behavior, but although these may change practices, there is no guaranteed positive effect on personal attitudes (35). Without changes in social norms, people often revert to old ways when incentives end or regulations are no longer enforced, and so long-term protection may be compromised.

Social capital can help to ensure compliance with rules and keep down monitoring costs, provided networks are dense, with frequent communication and reciprocal arrangements, small group size, and lack of easy exit options for members.

However, factors relating to the natural resources themselves—particularly when they are stationary, have high storage capacity (potential for biological growth), and clear boundaries—will also play a critical role in affecting whether social groups can succeed, keeping down the costs of enforcement, and ensuring positive resource outcomes (37).

Communities also do not always have the knowledge to appreciate that what they are doing may be harmful. For instance, it is common for fishing communities to believe that fish stocks are not being eroded, even though the scientific evidence indicates otherwise. Local groups may need the support of higher-level authorities, for example with legal structures that give communities clear entitlement to land and other resources as well as insulation from the pressures of global markets (8, 9). For global environmental problems, such as climate change, governments may need to regulate, partly because no community feels it can have a perceptible impact on a global problem. Thus, effective international institutions are needed to complement local ones (38).

Nonetheless, the ideas of social capital and governance of the commons, combined with the recent successes of local groups, offer routes for constructive and sustainable outcomes for natural resources in many of the world's ecosystems. To date, however, the triumphs of the commons have been largely at local to regional levels, where resources can be closed-access and where institutional conditions and market pressures are supportive. The greater challenge will center on applying some of these principles to open-access commons and worldwide environmental threats and on creating the conditions by which social capital can work under growing economic globalization.

References and Notes

1. G. Hardin, *Science* 162, 1243 (1968).
2. R. Nash, *Wilderness and the American Mind* (Yale Univ. Press, New Haven, 1973).
3. J. B. Callicott, M. P. Nelson, Eds., *The Great New Wilderness Debate* (Univ. of Georgia Press, Athens, 1998).
4. J. Pretty, *Agri-Culture: Reconnecting People, Land and Nature* (Earthscan, London, 2002).

5. D. Posey, Ed., *Cultural and Spiritual Values of Biodiversity* (IT Publishing, London, 1999).

6. M. Gadgil, R. Guha, *This Fissured Land: An Ecological History of India* (Oxford Univ. Press, New Delhi, 1992).

7. T. Beck, C. Naismith, *World Development* 29 (no. 1), 119 (2001).

8. E. Ostrom, *Governing the Commons* (Cambridge Univ. Press, New York, 1990).

9. E. Ostrom *et al.*, Eds., *The Drama of the Commons* (National Academy Press, Washington, DC, 2002).

10. S. Singleton, M. Taylor, *J. Theoret. Politics* 4, 309 (1992).

11. T. O'Riordan, S. Stoll-Kleeman, *Biodiversity, Sustainability and Human Communities* (Earthscan, London, 2002).

12. J. Dryzek, *Deliberative Democracy and Beyond* (Oxford Univ. Press, Oxford, 2000).

13. N. Uphoff, Ed., *Agroecological Innovations* (Earthscan, London, 2002).

14. J. Coleman, *Am. J. Sociol.* 94, S95 (1988).

15. R. D. Putnam, *Making Democracy Work* (Princeton Univ. Press, Princeton, NJ, 1993).

16. R. G. Wilkinson, *Ann. N.Y. Acad. Sci.* 896, 48 (1999).

17. R. Putnam, *Bowling Alone* (Simon & Schuster, New York, 2000).

18. J. Pretty, H. Ward, *World Dev.* 29 (no. 2), 209 (2001).

19. A. Agrawal, in *The Drama of the Commons*, E. Ostrom *et al.*, Eds. (National Academy Press, Washington, DC, 2002).

20. R Wade, *Village Republics* (ICS Press, San Francisco, ed. 2, 1994).

21. M. Taylor, *Community, Anarchy and Liberty* (Cambridge Univ. Press, Cambridge, 1982).

22. M. Woolcock, *Can. J. Policy Res.* 2, 11 (2001).

23. D. Narayan, L. Pritchett, *Cents and Sociability: Household Income and Social Capital in Rural Tanzania* (Policy Research Working Paper 1796, World Bank, Washington, DC, 1996).

24. A. Krishna, *Active Social Capital: Tracing the Roots of Development and Democracy* (Columbia Univ. Press, New York, 2002).

25. B. Wu, J. Pretty, *Agric. Human Values* 21, 81 (2004).

26. D. Pevalin, D. Rose, *Social Capital and Health* (Institute for Economic and Social Research, University of Essex, UK, 2003).

27. Fukuyama F. *Trust: The Social Values and the Creation of Prosperity* (New York: Free Press, 1995).

28. R. Putnam, *J. Democracy* 6, 65 (1995).

29. J. Knight, *Institutions and Social Conflict* (Cambridge Univ. Press, Cambridge, 1992).

30. A. Krishna, *Active Social Capital* (Columbia Univ. Press, New York, 2002).

31. C. B. Flora, J. L. Flora, *Am. Acad. Political Soc. Sci.* 529, 48 (1993).

32. K. S. Murali, I. K. Murthy, N. H. Ravindranath, *Environ. Manage. Health* 13, 512 (2002).

33. J. Pontius, R. Dilts, A. Bartlett, *From Farmer Field Schools to Community IPM* (FAO, Bangkok, 2001).

34. See the following Web sites for more data and evaluations on the ecological and economic impact of local groups: (i) Sustainable agriculture projects— analysis of 208 projects in developing countries in which social capital formation was a critical prerequisite of success, see www2.essex.ac.uk/ces/Research Programmes/subheads 4foodprodinc.htm. See also (*39*). (ii) Joint forest management (JFM) projects in India. For impacts in Andhra Pradesh, including satellite photographs, see www.ap.nic.in/apforest /jfm.htm. For case studies of JFM, see www.teriin.org/ jfm/cs.htm and www.iifm.org/databank/jfm/jfm.html. See also (*32, 40*). (iii) For community IPM, see www .communityipm.org/ and (*33*). (iv) For impacts on economic success in rural communities, see (*41, 42*). (v) For Landcare program in Australia, where 4,500 groups have formed since 1989, see www.landcare australia.com.au/projectlist.asp and www.landcare australia.com.au/FarmingCaseStudies.asp.

35. G. T. Gardner, P. C. Stern, *Environmental Problems and Human Behavior* (Allyn and Bacon, Needham Heights, MA, 1996)

36. P. C. Stern, T. Dietz, N. Doscak, E. Ostrom, S. Stonich, in (*9*), pp. 443–490.

37. P. M. Haas, R. O. Keohane, M. A. Levy, Eds., in *Institutions for the Earth*, (MIT Press, Cambridge, 1993).

38. J. Pretty, J. I. L. Morison, R. E. Hine, *Agric. Ecosys. Environ.* 95 (1), 217 (2003).

39. K. S. Murali *et al.*, *Int. J. Environ. Sustainable Dev.* 2, 19 (2003).

40. D. Narayan, L. Pritchett, *Cents and Sociability: Household Income and Social Capital in Rural Tanzania*, World Bank Policy Research Working Paper 1796 (World Bank, Washington, DC, 1997). Available at: http://poverty.worldbank.org/library/view/6097/.

41. P. Donnelly-Roark, X. Ye, *Growth, Equity and Social Capital: How Local Level Institutions Reduce Poverty* (World Bank, Washington, DC, 2002). Available at: http://poverty.worldbank.org/library/view/13137.

Managing Tragedies
Understanding Conflict
over Common Pool Resources

WILLIAM M. ADAMS, DAN BROCKINGTON, JANE DYSON,
and BHASKAR VIRA

onflicts over the management of common pool resources are not simply material. They also depend on the perceptions of the protagonists. Policy to improve management often assumes that problems are self-evident, but in fact careful and transparent consideration of the ways different stakeholders understand management problems is essential to effective dialogue.

The management of common pool resources can be viewed as a problem of collective action and can be analyzed in terms of the costs and benefits of cooperation, institutional development, and monitoring, according to variables such as group size, composition, relationship with external powers, and resource characteristics (1–5). However, resulting policy debates are often flawed because of the assumption that the

This article first appeared in *Science* (12 December 2003: Vol. 302, no. 5652). It has been revised and updated for this edition.

actors involved share an understanding of the problem that is being discussed. They tend to

KEY TERM

Common pool resources are goods from which an entity, such as a country or individual, cannot be excluded, and in which consumption by the beneficiary detracts from other users' ability to consume the good. Without well-established communal norms of use and/or management regimes, they are difficult to protect and easy to deplete. Common pool resources differ from open access resources in that the latter lack usage norms or regulation.

> The problem of fuel wood scarcity, as perceived by government planners and donors, was quite different from the problems of primary concern to small farmers, who were trying to secure access to sustainable livelihood options by optimizing the use of their land and labor resources.

ignore the fact that the assumptions, knowledge, and understandings that underlie the definition of resource problems are frequently uncertain and contested.

Recent policy debates over natural resource management have revealed the unexpected consequences of the assumption that problems (and therefore solutions) are self-evident. For example, in the 1970s and 1980s, governments and donors in Asia and sub-Saharan Africa perceived an impending fuel wood crisis. In response, they developed social forestry projects to persuade smallholders to plant trees on their farmlands, assuming that the aggregate assessment of shortages reflected an acute need for fuel wood at the household level. Subsequent research showed that the assumptions behind these interventions were deeply flawed (6, 7). Some households indeed planted trees to provide fuel. Many others responded to scarcity by sharing cooking arrangements, increasing labor devoted to collection, substituting between fuels, migrating, or engaging in nomadism and transhumance [*transhumance* is moving of livestock (cows, sheep, horses) from one environment to another in response to seasonal availability of forage]. Furthermore, those farmers who planted trees primarily did so as a cash crop to be sold for pulp, small timber, and poles, not as a subsistence com-

modity for fuel wood. The problem of fuel wood scarcity, as perceived by government planners and donors, was quite different from the problems of primary concern to small farmers, who were trying to secure access to sustainable livelihood options by optimizing the use of their land and labor resources.

Similarly, concern over appropriate rangeland use in Africa has long been dominated by the perception that pastures were being overstocked (8). A variety of self-evident notions supported this, including the idea that cattle were kept purely for prestige and that use of commonly held pastures could not be regulated.

Subsequent research into dry-land ecology, herd use, and pasture management suggests that high stocking rates can make sense where ecosystem productivity is driven by variable rainfall (9–11). Too many livestock will still cause problems for some herders, but there are no grounds for the unmitigated gloom surrounding policy debates about the overstocking of these rangelands.

Defining Resource Management Problems

Problem definition is critical to the process of making policy, yet its role is rarely scrutinized. Stakeholders often do not explicitly recognize the ways in which their knowledge and understanding frame their perspectives on common pool resource management policy. Although conflict is a feature of many resource management regimes, it is often assumed to reflect differences in material interests between stakeholders. In such circumstances, conflict may be managed by trading off different management objectives (12) or by attempting to reconcile multiple interests in resource management (13).

We suggest that the origins of conflict go beyond material incompatibilities. They arise at a deeper cognitive level. In our view, stakeholders draw on their current knowledge and understanding to cognitively frame a specific common pool resource management problem. Thus, differences in knowledge, understanding, preconceptions, and priorities are often obscured in conventional policy dialogue and may provide a deeper explanation of conflict.

It is precisely when different stakeholders (of different sizes and operating at different levels) reveal different interpretations of key issues that the policy debate can be most productive. The knowledge that allows stakeholders to define the problems of resource use falls into three realms: knowledge of the empirical context; knowledge of laws and institutions; and beliefs, myths, and ideas.

Stakeholders' knowledge of the empirical context derives from a variety of sources. At the local level, knowledge may derive from direct personal experience, particularly of extreme events such as droughts or floods. At larger scales, knowledge may reflect inference from known changes elsewhere, such as the insights of formal empirical and theoretical research by official agencies and research organizations using remote sensing, censuses, or sample surveys.

Decision makers and stakeholders are likely to differ in their access to, and understanding of, these diverse sources of knowledge. Thus, the knowledge of any particular stakeholder will be partial and hence may be contested by other actors. The ideas, ideologies, and beliefs brought to bear on problem definition by different stakeholders can substantially influence problem perception. Religious beliefs and moral conviction can be important in structuring understanding, both among local people and scientists. Ideas derive legitimacy from received wisdom about theory [e.g., the widespread belief in the "Tragedy of the Commons" (14); see also the chapter "The Tragedy of the Commons" in this volume] and from ideas outside formal science, including informal or "folk" knowledge.

Policy narratives or story lines can exert a powerful influence on official decision makers' perceptions of resource management problems (15–17). Divergent received wisdoms used by different actors in their analysis are a potent source of conflict over appropriate response options.

RANDOM SAMPLES

Science, Vol. 307, no. 5715, 1558,
11 March 2005

MAN THE ERODER

For the past millennium, humans have been moving more earth than all natural processes combined. Just how far have we tipped the balance? Geologist Bruce Wilkinson of the University of Michigan, Ann Arbor, decided to find out.

He calculated prehistoric rates of erosion through the amount of sedimentary rock, the end result of erosion, that has accumulated over the past 500 million years and estimated that natural erosion lowers Earth's land surface about 24 meters every million years. He then calculated the human contribution, combining estimates of erosion from crop tillage, land conversion for grazing, and construction. Averaged out over the world's land surface, that came to about 360 meters per million years, or 15 times the natural rate.

This difference amply demonstrates that current agricultural practices are unsustainable, says Wilkinson, who points out in the March 2005 issue of *Geology* that at the current rate, the soil eroded from Earth's surface would fill the Grand Canyon in 50 years.

Wilkinson's estimates for natural erosion are similar to those of geologist Paul Bierman of the University of Vermont, Burlington, who has used beryllium isotopes to estimate erosion rates in the Appalachian Mountains from the past 10,000 to 100,000 years. "The most intriguing part of this study is to be able to look back over 500 million years of earth history," says Bierman.

Most common pool resource management situations do not operate in isolation from a wider context of the legal and institutional framework. Stakeholders differ in their knowledge of this framework: A local herder may be unaware of a country's policy commitments under the Convention on Biological Diversity, whereas a state resource manager may be forced to act in particular ways because of commitments under such multilateral agreements. In this sense, knowledge about laws and institutions may be seen as providing both constraints and opportunities for common pool resource management, because this knowledge forces stakeholders to consider resource uses that are compatible with these wider policy processes. Importantly, knowledge about policy is likely to contribute to the way in which a stakeholder perceives a problem and hence the alternative responses that he or she is willing to consider.

Cognitive Conflict in Common Pool Resource Management

If each stakeholder is able to define problems and test the set of possible response options only in the context of his or her particular knowledge and

> . . . a market may develop for something previously regarded locally as useless or destructive of value, such as wildlife tourism. In these situations, the realm of conflict between beneficiaries and others will be both cognitive and material.

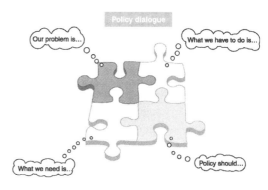

FIGURE 25.   Cognitive conflict over common pool resources.

understanding, then agreement is less likely, both in terms of perceptions and problem definition and in terms of the desired response to the problem. Thus, policy conflict arises because differences in knowledge and understanding between stakeholders frame their perceptions of resource use problems as well as possible solutions to these problems (Figure 25).

One cannot, therefore, simply analyze the economic interests of different claimants to rights over a defined resource.

Different people will see different resources in a landscape. They will perceive different procedures appropriate for reconciling conflict. Moreover, perceptions will change, because different elements within the landscape will become "resources." For example, a market may develop for something previously regarded locally as useless or destructive of value, such as wildlife tourism. In these situations, the realm of conflict between beneficiaries and others will be both cognitive and material.

Where cognitive conflict is important, policy dialogue needs to be structured so that differences in knowledge, understanding, ideas, and beliefs in the public arena are recognized. Ostrom's seminal studies of the evolution of institutions of water management in California demonstrate the value of precisely this dialogue, and the absence of dialogue explains the failure of these institutions to emerge (18). Similarly, Wilson has argued that only by making explicit the uncertainties and ignorance inherent in understanding fisheries will it be possible to establish the collective learning experience necessary to manage resource use in such complex ecosystems (14).

Making explicit the basis of different stakeholders' positions is likely to improve the transparency and effectiveness of negotiations between stakeholders by enabling actors to understand the plurality of views that prevail in the context of resource use and management. Failure to recognize the cognitive dimension of conflict results in superficial policy measures that fall short of addressing the deeper underlying (structural) differences between resource users.

Of course, a deeper understanding of stakeholder differences over common pool resources does not guarantee that policy negotiations will result in win-win scenarios, but it may smooth the path toward consensus in situations where there are incompatibilities in stakeholders' interests, values, or priorities. Management effectiveness will always be limited by incomplete knowledge and understanding of complex natural and social systems. Our type of reasoning will not help if decisions are driven by the unilateral political will or the economic power of particular stakeholders. If policy is made in a way that precludes dialogue, our approach will be of limited use, except to explain why things go wrong.

This perspective on resource conflict can reveal the incompatibility of competing perceptions, but it cannot by itself reconcile them. Techniques for conflict resolution, negotiation, and management must then come into play. To some extent, policy will always involve "tragic" choices that contradict the deeply held values and beliefs of some stakeholders (19).

References and Notes

1. A. Agrawal, *World Dev.* 29, 1648 (2001).
2. R. J. Oakerson, in *Proceedings of the Conference on Common Property Resource Management*, National Research Council, Ed. (National Academies Press, Washington, DC, 1986), pp. 13–30.
3. R. J. Oakerson, in *Making the Commons Work: Theory, Practice and Policy*, D. W. Bromley, Ed. (Institute for Contemporary Studies Press, San Francisco, CA, 1992), pp. 41–59.
4. J. T. Thomson, D. Feeny, R. J. Oakerson, in *Making the Commons Work: Theory, Practice and Policy*, D. W. Bromley, Ed. (Institute for Contemporary Studies Press, San Francisco, CA, 1992), pp. 129–160.
5. V. M. Edwards, N. A. Steins, *J. Theor. Polit.* 10, 347 (1998).
6. J. E. M. Arnold, P. Dewees, *Farms, Trees and Farmers: Responses to Agricultural Intensification* (Earthscan, London, 1997).
7. G. Leach, R. Mearns, *Beyond the Woodfuel Crisis: People, Land and Trees in Africa* (Earthscan, London, 1988).
8. A. R. E. Sinclair, J. M. Fryxell, *Can. J. Zool.* 63, 987 (1985).
9. R. H. Behnke, I. Scoones, in *Range Ecology at Disequilibrium: New Models of Natural Variability and Pastoral Adaptation in African Savannas*, R. H. Behnke, I. Scoones, C. Kerven, Eds. (Overseas Development Institute, London, 1993), pp. 1–30.
10. S. Sullivan, R. Rohde, *J. Biogeogr.* 29, 1595 (2002).
11. A. Illius, T. O'Connor, *Ecol. Appl.* 9, 798 (1999).
12. K. Brown, W. N. Adger, E. Tompkins, P. Bacon, D. Shim, K. Young, *Ecol. Econ.* 37, 417 (2001).
13. E. Wollenberg, J. Anderson, D. Edmunds, *Int. J. Agric. Resourc. Governance Ecol.* 1, 199 (2001).
14. J. Wilson, in *The Drama of the Commons*, E. Ostrom *et al.*, Eds. (National Academies Press, Washington DC, 2002), pp. 327–359.
15. E. Roe, *World Dev.* 19, 287 (1991).
16. M. Leach, R. Mearns, in *The Lie of the Land. Challenging Received Wisdom on the African Environment*, M. Leach, R. Mearns, Eds. (James Currey, Oxford, 1996), pp. 1–33.
17. J. Dryzek, *The Politics of the Earth: Environmental Discourses* (Oxford Univ. Press, London, 1997).
18. E. Ostrom, *Governing the Commons. The Evolution of Institutions for Collective Action* (Cambridge Univ. Press, Cambridge, 1990).
19. G. Calabresi, P. Bobbitt, *Tragic Choices* (Norton, New York, 1978).
20. This paper draws on research funded by the UK Department for International Development (DFID) Natural Resources Systems Programme (project R7973). We thank K. Chopra, M. Murphree, and I. Shivji, who shared this project. The views expressed are not necessarily those of DFID.

Global Food Security
Challenges and Policies

MARK W. ROSEGRANT and SARAH A. CLINE

Global food security will remain a worldwide concern for the next 50 years and beyond. Recently, crop yield has fallen in many areas because of declining investments in research and infrastructure, as well as increasing water scarcity. Climate change and HIV/AIDS are also crucial factors affecting food security in many regions. Although agroecological approaches offer some promise for improving yields, **food security** in developing countries could be substantially improved by increased investment and policy reforms.

The ability of agriculture to support growing populations has been a concern for generations and continues to be high on the global policy agenda. The eradication of poverty and hunger was included as one of the United Nations Millennium Development Goals adopted in 2000. One of the targets of the Goals is to halve the proportion of people who suffer from

This article first appeared in *Science* (12 December 2003: Vol. 302, no. 5652). It has been revised and updated for this edition.

> ## KEY TERM
>
> **Food security** most commonly refers to the situation in which all people at all times have both physical and economic access to sufficient food to meet their dietary needs for a productive and healthy life. The term is used in development and humanitarian aid fields, in which the focus of food security interventions is usually the development of indigenous coping mechanisms to fight hunger and malnutrition.

hunger between 1990 and 2015 (*1*). Meeting this food security goal will be a major challenge. Predictions of food security outcomes have been a part of the policy landscape since Malthus's *An Essay on the Principle of Population* of 1798 (*2*). Over the past several decades, some experts have expressed concern about the ability of agricultural production to keep up with global food demands (*3–5*), whereas others have forecast that technological advances or expansions of cultivated area would boost production sufficiently to meet rising demands (*6–8*). So far, dire predictions of a global food security catastrophe have been unfounded.

Nevertheless, crop yield growth has slowed in much of the world because of declining investments in agricultural research, irrigation, and rural infrastructure and increasing water scarcity. New challenges to food security are posed by climate change and the morbidity and mortality of human immunodeficiency virus/acquired immunodeficiency syndrome (HIV/AIDS). Many studies predict that world food supply will not be adversely affected by moderate climate change, assuming that farmers will take adequate steps to adjust to climate change and that additional CO_2 will increase yields (*9*).

However, many developing countries are likely to fare badly. In warmer or tropical environments, climate change may result in more intense rainfall events between prolonged dry periods, as well as reduced or more variable water resources for irrigation. Such conditions may promote pests and disease on crops and livestock, as well as soil erosion and desertification. Increasing development into marginal lands may in turn put these areas at greater risk of environmental degradation (*10, 11*). The HIV/AIDS epidemic is another global concern, with an estimated 42 million cases worldwide at the end of 2002 (*12*); 95 percent of those are in developing countries. In addition to its direct health, economic, and social impacts, the disease also affects food security and nutrition. Adult labor is often removed from affected households, and these households will have less capacity to produce or

buy food, as assets are often depleted for medical or funeral costs (*13*). The agricultural knowledge base will deteriorate as individuals with farming and science experience succumb to the disease (*14*). Can food security goals be met in the face of these old and new challenges?

Several organizations have developed quantitative models that project global food supply and demand into the future (*15–19*). According to the most recent baseline projections of the International Food Policy Research Institute's (IFPRI's) International Model for Policy Analysis of Agricultural Commodities and Trade (IMPACT) (*20*), global cereal production is estimated to increase by 56 percent between 1997 and 2050, and livestock production by 90 percent. Developing countries will account for 93 percent of cereal-demand growth and 85 percent of meat-demand growth to 2050. Income growth and rapid urbanization are major forces driving increased demand for higher valued commodities, such as meats, fruits, and vegetables. International agricultural trade will increase substantially, with developing countries' cereal imports doubling by 2025 and tripling by 2050. Child malnutrition will persist in many developing countries, although overall, the share of malnourished children is projected to decline from 31 percent in 1997 to 14 percent in 2050 (Figure 26). Nevertheless, this represents a nearly 35-year delay in meeting the Millennium Development Goals. In some places, circumstances will deteriorate, and in sub-Saharan Africa, the number of malnourished preschool children will increase between 1997 and 2015, after which they will only decrease slightly until 2050. South Asia is another region of concern—although progress is expected in this region, more than 30 percent of preschool children will remain malnourished by 2030, and 24 percent by 2050 (*21*).

Achieving food security needs policy and investment reforms on multiple fronts, including human resources, agricultural research, rural infrastructure, water resources, and farm- and community-based agricultural and natural resources management. Progressive policy

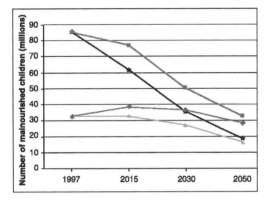

FIGURE 26. Projected number of malnourished children in South Asia and sub-Saharan Africa under baseline and progressive policy actions scenarios, 1997 to 2050. Blue, South Asia, baseline scenario; black, South Asia, progressive policy actions scenario; dark gray, sub-Saharan Africa, baseline scenario; light gray, sub-Saharan Africa, progressive policy actions scenario. *Source:* International Food Policy Research Institute (*21*).

action must not only increase agricultural production, but it must also boost incomes and reduce poverty in rural areas where most of the poor live. If we take such an approach, we can expect production between 1997 and 2050 to increase by 71 percent for cereals and by 131 percent for meats. A reduction in childhood malnutrition would follow; the number of malnourished children would decline from 33 million in 1997 to 16 million in 2050 in sub-Saharan Africa, and from 85 million to 19 million in South Asia (Figure 26).

Increased investment in people is essential to accelerate food security improvements. In agricultural areas, education works directly to enhance the ability of farmers to adopt more advanced technologies and crop-management techniques and to achieve higher rates of return on land (*22*). Moreover, education encourages movement into more remunerative nonfarm work, thus increasing household income. Women's education affects nearly every dimension of development, from lowering fertility rates

to raising productivity and improving environmental management (*23*). Research in Brazil shows that 25 percent of children were stunted if their mothers had four or fewer years of schooling; however, this figure fell to 15 percent if the mothers had a primary education and to 3 percent if mothers had any secondary education (*24*). Poverty reduction is usually enhanced by an increase in the proportion of educational resources going to primary education and to the poorest groups or regions (*25–27*). Investments in health and nutrition, including safe drinking water, improved sewage disposal, immunization, and public health services, also contribute to poverty reduction. For example, a study in Ethiopia shows that the distance to a water source, as well as nutrition and morbidity, all affect agricultural productivity of households (*28*).

When rural infrastructure has deteriorated or is nonexistent, the cost of marketing farm produce—thus escaping subsistence agriculture and improving incomes—can be prohibitive for poor farmers. Rural roads increase agricultural production by bringing new land into cultivation and by intensifying existing land use, as well as consolidating the links between agricultural and nonagricultural activities within rural areas and between rural and urban areas (*29*). Government expenditure on roads is the most important factor in poverty alleviation in rural areas of India and China, because it leads to new employment opportunities, higher wages, and increased productivity (*30, 31*).

In addition to being a primary source of crop and livestock improvement, investment in agricultural research has high economic rates of return (*32*). Three major yield-enhancing strategies include research to increase the harvest index (*33*), plant biomass, and stress tolerance (particularly drought resistance) (*34, 35*). For example, the hybrid "New Rice for Africa," which was bred to grow in the uplands of West Africa, produces more than 50 percent more grain than current varieties when cultivated in traditional rainfed systems without fertilizer. Moreover, this variety matures 30 to 50 days earlier than current varieties and has enhanced disease and drought tolerance (36). In addition to conventional breeding, recent developments in nonconventional breeding, such as marker-assisted selection and cell and tissue culture techniques, could be employed for crops in developing countries, even if these countries stop short of transgenic breeding. To date, however, the application of molecular biotechnology has been mostly limited to a small number of traits of interest to commercial farmers, mainly developed by a few global life science companies.

Although much of the science and many of the tools and intermediate products of biotechnology are transferable to solve high-priority problems in the tropics and subtropics, it is generally agreed that the private sector will not invest sufficiently to make the needed adaptations in these regions with limited market potential. Consequently, the public sector will have to play a key role, much of it by accessing proprietary tools and products from the private sector (37).

Irrigation is the largest water user worldwide, but also the first sector to lose out as scarcity increases (38). The challenges of water scarcity are heightened by the increasing costs of developing new water sources, soil degradation in irrigated areas, groundwater depletion, water pollution, and ecosystem degradation. Wasteful use of already-developed water supplies may be encouraged by subsidies and distorted incentives that influence water use. Hence, investment is needed to develop new water management policies and infrastructure. Although the economic and environmental costs of irrigation make many investments unprofitable, much could be achieved by water conservation and increased efficiency in existing systems and by increased crop productivity per unit of water used. Regardless, more research and policy efforts need to be focused on rainfed agriculture. Exploiting the full potential of rainfed agriculture will require investment in water harvesting technologies, crop breeding, and extension services, as well as good access to markets, credit, and supplies. Water harvesting and conservation tech-

> ## KEY TERM
>
> Conventional breeding develops new plant varieties by the process of selection, and seeks to achieve expression of genetic material which is already present within a species. Nonconventional breeding, or genetic engineering, works primarily through insertion of new genetic material not present in a species, although gene insertion must also be followed up by selection.

niques are particularly promising for the semi-arid tropics of Asia and Africa, where agricultural growth has been less than 1 percent in recent years. For example, water harvesting trials in Burkina Faso, Kenya, Niger, Sudan, and Tanzania show increases in yield of a factor of two to three, compared with dryland farming systems (39, 40).

Agroecological approaches that seek to manage landscapes for both agricultural production and ecosystem services are another way of improving agricultural productivity. A study of 45 projects, using agroecological approaches, in 17 African countries shows cereal yield improvements of 50 to 100 percent (41). There are many concomitant benefits to such approaches, as they reduce pollution through alternative methods of nutrient and pest management, create biodiversity reserves, and enhance habitat quality through careful management of soil, water, and natural vegetation. Important issues remain about how to scale up agroecological approaches. Pilot programs are needed to work out how to mobilize private investment and to develop systems for payment of ecosystem services. All of these issues require investment in research, system development, and knowledge sharing.

To implement agricultural innovation, we need collective action at the local level, as well as the participation of government and nongovernmental organizations that work at the community level. There have been several successful programs, including those that use water harvesting and conservation techniques (42, 43). Another priority is participatory plant breeding for yield increases in rainfed agrosystems, particularly in dry and remote areas. Farmer participation can be used in the very early stages of breed selection to help find crops suited to a multitude of environments and farmer preferences. It may be the only feasible route for crop breeding in remote regions, where a high level of crop diversity is required within the same farm, or for minor

SCIENCE IN THE NEWS

Science, Vol. 288, no. 5465, 429, 21 April 2000

HOPES GROW FOR HYBRID RICE TO FEED DEVELOPING WORLD
Dennis Normile

LOS BAÑOS, THE PHILIPPINES—Sant Virmani, who heads hybrid rice-breeding efforts at the International Rice Research Institute (IRRI) here, remembers when the number of scientists interested in the subject could fit into his living room. But this month, organizers of an international conference marking the 40th anniversary of IRRI had to fold back a room divider and bring in more chairs to handle the throng that gathered to hear his talk.

Hopes for Hybrids: Hybrid rice varieties could mean hardier crops, higher yields, and more food to feed the rapidly expanding population of the third world.
CREDIT: PAUL PIEBINGA/ISTOCKPHOTO.COM

That heightened interest reflects the growing number of researchers who hope that hybrid rice will help feed the billions of people who rely on the crop. "Hybrid rice is really the only [technique] at hand that has proven to boost yields in farmers' fields," Virmani says.

Although rice breeders have created improved, higher-producing rice varieties, they haven't been able to take advantage of a natural phenomenon that jacks up the yields of grains such as corn. Thanks to an imperfectly understood effect called heterosis, the first generation, or F1, hybrid of a cross of two different varieties grows more vigorously and produces from 15 percent to 30 percent more grain than either parent. But because rice is self-pollinating, with each plant producing its own fertilizing pollen, producing hybrid rice was commercially impractical. Now, three decades of effort has produced hybrid rice varieties and commercially viable methods of producing the hybrid seed. "Finally, hybrid rice is ready to take off," Virmani says.

Such a jump is needed because increases in rice yields have leveled off in the 1990s, while the population continues to grow. But others counsel caution. They warn that the quality of the hybrid rice hasn't yet matched that of current varieties and that growing hybrid rice requires changes in farming practices, in particular, the purchase of new seeds for every growing season. "Hybrid rice is not a success story—yet," says Wayne Freeman, a retired agronomist who formerly oversaw The Rockefeller Foundation's food programs in India.

crops that are neglected by formal breeding programs (*44, 45*).

Making substantial progress in improving food security will be difficult, and it does mean reforming currently accepted agricultural practices. However, innovations in agroecological approaches and crop breeding have brought some documented successes. Together with investment in research and water and transport infrastructure, we can make major improvements to global food security, especially for the rural poor.

References and Notes

1. The World Bank Group, "Millennium Development Goals: About the Goals," www.developmentgoals.org/About_the_goals.htm (2003).

2. T. R. Malthus, *An Essay on the Principle of Population* (Norton, New York, 2003).

3. P. R. Ehrlich, A. H. Ehrlich, *The Population Explosion* (Simon & Schuster, New York, 1990).

4. L. Brown, H. Kane, *Full House: Reassessing the Earth's Population Carrying Capacity* (Norton, New York, 1994).

5. D. H. Meadows, D. L. Meadows, J. Randers, *Beyond the Limits: Confronting Global Collapse, Envisioning a Sustainable Future* (Chelsea Green, White River Junction, VT, 1992).

6. E. Boserup, *The Conditions of Agricultural Growth: The Economics of Agrarian Change Under Population Pressure* (Aldine, Chicago, 1965).

7. C. Clark, *Starvation or Plenty* (Taplinger, New York, 1970).

8. J. L. Simon, *The Ultimate Resource 2* (Princeton Univ. Press, Princeton, NJ, 1998).

9. R. M. Adams, B. H. Hurd, *Choices* 14, 22 (1999).

10. T. E. Downing, *Renewable Energy* 3, 491 (1993).

11. International Panel on Climate Change, *Climate Change 2001: Impacts, Adaptation, and Vulnerability* (Cambridge Univ. Press, Cambridge, 2001).

12. Joint United Nations Program on HIV/AIDS (UNAIDS), www.unaids.org (2003).

13. Food and Agricultural Organization of the United Nations (FAO), "The impact of HIV/AIDS on food security," Committee on World Food Security, 27th Session, Rome, 28 May to 1 June, 2001.

14. L. Haddad, S. Gillespie, *J. Int. Dev.* 13, 487 (2001).

15. J. Bruinsma, Ed., *World Agriculture: Towards 2015/ 2030: An FAO Study* (Earthscan, London, 2003).

16. Food and Agricultural Policy Research Institute, *U.S. and World Agricultural Outlook 2003* (Staff Report 1-03, Food and Agricultural Policy Research Institute,

University of Iowa and University of Missouri– Columbia, 2003).

17. M. W. Rosegrant *et al.*, *Global Food Projections to 2020: Emerging Trends and Alternative Futures* (IFPRI, Washington, DC, 2001).

18. Organisation for Economic Co-operation and Development, *Agricultural Outlook 2003–2008* (Organisation for Economic Co-operation and Development, Paris, 2003).

19. U.S. Department of Agriculture, Office of the Chief Economist, *USDA Agricultural Baseline Projections to 2012* (Staff Report WAOB-2003-1, U.S. Department of Agriculture, Washington, DC, 2003).

20. M. W. Rosegrant, S. Meijer, S. A. Cline, "International Model for Policy Analysis of Agricultural Commodities and Trade (IMPACT): Model description" (IFPRI, Washington, DC, 2002), available at www.ifpri.org/themes/impact/impactmodel.pdf.

21. International Food Policy Research Institute, unpublished data.

22. World Bank, *Poverty: World Development Indicators* (Oxford Univ. Press, New York, 1990).

23. World Bank, *Poverty Reduction and the World Bank: Progress and Challenges in the 1990s* (The International Bank for Reconstruction and Development/ The World Bank, Washington, DC, 1996).

24. A. Kassouf, B. Senauer, *Econ. Dev. Cult. Change* 44, 817 (1996).

25. M. Lipton, S. Yaqub, E. Darbellay, *Success in Anti-Poverty* (International Labour Office, Geneva, 1998).

26. R. Gaiha, *Design of Poverty Alleviation Strategy in Rural Areas* (FAO, Rome, 1994).

27. R. Singh, P. Hazell, *Econ. Polit. Wkly.* 28, A9 (20 March 1993).

28. A. Croppenstedt, C. Muller, *Econ. Dev. Cult. Change* 48, 475 (2000).

29. International Fund for Agricultural Development (IFAD), *The State of World Rural Poverty: A Profile of Asia* (IFAD, Rome, 1995).

30. S. Fan, P. Hazell, S. Thorat, *Am. J. Agric. Econ.* 82, 1038 (2000).

31. S. Fan, L. Zhang, X. Zhang, "Growth, inequality, and poverty in rural China: The role of public investments" (IFPRI Research Report 125, IFPRI, Washington, DC, 2002).

32. J. M. Alston, P. G. Pardey, C. Chan-Kang, T. J. Wyatt, M. C. Marra, "A meta-analysis of rates of return to agricultural R&D: Ex pede Herculem?" (IFPRI Research Report 113, IFPRI, Washington, DC, 2000).

33. The harvest index is defined as "the ratio of grain to total crop biomass" (*34*).

34. K. G. Cassman, *Proc. Natl. Acad. Sci. U.S.A.* 96, 5952 (1999).

35. L. T. Evans, *Feeding the Ten Billion: Plants and Population Growth* (Cambridge Univ. Press, Cambridge, 1998).

36. West Africa Rice Development Association, "Consortium formed to rapidly disseminate New Rice for Africa," available at www.warda.org/warda1/main/newsrelease/newsrel-consortiumapro1.htm (cited 2000).

37. D. Byerlee, K. Fischer, "Accessing modern science: Policy and institutional options for agricultural biotechnology in developing countries" (Agricultural Knowledge and Information Systems Discussion Paper, The World Bank, Washington, DC, 2000).

38. M. W. Rosegrant, X. Cai, S. A. Cline, *World Water and Food to 2025: Dealing with Scarcity* (IFPRI, Washington, DC, 2002).

39. S. M. Barghouti, *Sustain. Dev. Int.* 1, 127 (2001), www.sustdev.org/agriculture/articles/edition1/01.127.pdf.

40. Food and Agriculture Organization of the United Nations, *Crops and Drops* (FAO, Rome, 2000).

41. J. Pretty, *Environ. Dev. Sustain.* 1, 253 (1999).

42. J. Pretty, *World Dev.* 23, 1247 (1995).

43. T. Oweis, A. Hachum, J. Kijne, "Water harvesting and supplementary irrigation for improved water use efficiency in dry areas" (System-Wide Initiative on Water Management Paper 7, International Water Management Institute, Colombo, Sri Lanka, 1999).

44. S. Ceccarelli, S. Grando, R. H. Booth, in *Participatory Plant Breeding*, P. Eyzaguire, M. Iwanaga, Eds. (International Plant Genetics Research Institute, Rome, 1996), pp. 99–116.

45. J. Kornegay, J. A. Beltran, J. Ashby, in *Participatory Plant Breeding*, P. Eyzaguire, M. Iwanaga, Eds. (International Plant Genetics Research Institute, Rome, 1996), pp. 151–159.

New Visions for Addressing Sustainability

ANTHONY J. McMICHAEL, COLIN D. BUTLER, and CARL FOLKE

Attaining sustainability will require concerted interactive efforts among disciplines, many of which have not yet recognized, and internalized, the relevance of environmental issues to their main intellectual discourse. The inability of key scientific disciplines to engage interactively is an obstacle to the actual attainment of sustainability. For example, in the list of Millennium Development Goals from the United Nations World Summit on Sustainable Development, Johannesburg, 2002, the seventh of the eight goals—to "ensure environmental sustainability"—is presented separately from the parallel goals of reducing fertility and poverty, improving gains in equity, improving material conditions, and enhancing population health. A more integrated and consilient approach to sustainability is urgently needed.

This article first appeared in *Science* (12 December 2003: Vol. 302, no. 5652). It has been revised and updated for this edition.

For human populations, sustainability means transforming our ways of living to maximize the chances that environmental and social conditions will indefinitely support human security, well-being, and health. In particular, the flow of nonsubstitutable goods and services from

KEY TERM

Most goods have substitutes— for example, margarine can be substituted for butter. However, some goods, such as water, are unique and essential and have no economic substitutes. In economics, once a good becomes too costly, the cost will drive innovation to find a lower-cost substitute. This is not the case for **nonsubstitutable goods**.

SCIENCE IN THE NEWS

Science, Vol. 287, no. 5456, 1192–1195, 18 February 2000

A NEW BREED OF SCIENTIST-ADVOCATE EMERGES
Kathryn S. Brown

For months, David Wilcove peppered the U.S. Fish and Wildlife Service (FWS) with letters protesting the agency's plans to save the threatened Utah prairie dog. Wilcove, a conservation biologist, and his colleagues at Environmental Defense in Washington, D.C., argued that FWS was putting too much emphasis on protecting prairie dogs on federal lands, when most of the animals now live on private land and cannot be relocated easily.

In the midst of this typical conservation battle—scientist-advocates on one side, resource managers on the other—Wilcove made an atypical move. Conceding that his organization and the FWS were both shooting from the hip, making cases based on skimpy data, he flew a team from Princeton University to Utah to meet with agency managers and Environmental Defense officials. The Princeton group, led by biologist Andrew Dobson, began working up what the cash-strapped FWS could not afford to do on its own: a model on how various factors, from climate to disease epidemics, would affect Utah prairie dogs. "When the study is done, we'll all have a better blueprint for determining the relative importance of public and private lands," Wilcove says.

That kind of cooperation is a novel way to get more science into resource management decisions. Week in and week out, managers dictate which sections of forest to sell to logging companies, which wetlands to pave over for houses, and which prairies to till into pastures. Such decisions often are justified by price tag or politics, but it's rare that more than lip service is paid to science. Part of the problem is that many scientists are hesitant, or unable, to participate in the process. "Academics don't know how to affect policy, and they don't communicate with managers very well," says Michael Soulé, a professor emeritus at the University of California, Santa Cruz.

ecosystems must be sustained. The contemporary stimulus for exploring sustainability is the accruing evidence that humankind is jeopardizing its own longer-term interests by living beyond Earth's means, thereby changing atmospheric composition and depleting biodiversity, soil fertility, ocean fisheries, and freshwater supplies (1).

Much early discussion about sustainability has focused on readily measurable intermediate outcomes such as increased economic performance, greater energy efficiency, better urban design, improved transport systems, better conservation of recreational amenities, and so on. However, such changes in technologies, behaviors, amenities, and equity are only the means to attaining desired human experiential outcomes,

including autonomy, opportunity, security, and health. These are the true ends of sustainability—and there has been some recognition that their attainment, and their sharing, will be optimized by reducing the rich-poor divide (2).

Some reasons for the failure to achieve a collective vision of how to attain sustainability lie in the limitations of, and disjunction between, disciplines we think should be central to our understanding of sustainability: demography, economics, ecology, and epidemiology. These disciplines bear on the size and economic activities of the human population, how the population relates to the natural world, and the health consequences of ecologically injudicious behavior. Sustainability issues are of course not limited to these four disciplines, but require the engagement and interdisciplinary collaboration of other social and natural sciences, engineering, and the humanities (3).

Neither mainstream demography nor economics, for the most part, incorporate sufficient appreciation of environmental criticalities into their thinking. They implicitly assume that the world is a quasi steady-state system within which discipline-specific processes can be studied. Although contemporary ecology has broadened its perspectives significantly, there is still a tendency to insufficiently consider both human influence and dependence on ecosystem composition, development, and dynamics. Epidemiologists focus mainly on individual-level behaviors and circumstances as causes of disease. This discounts the underlying social, cultural, and political determinants of the distribution of disease risk within and between populations, and has barely recognized the health risks posed by today's global environmental changes.

These four disciplines share a limited ability to appreciate that the fate of human populations depends on the biosphere's capacity to provide a continued flow of goods and services. The assumption of human separateness from the natural world perpetuates a long-standing, biblically based premise of Western scientific thought of Man as master, with dominion over Nature (4).

Many disciplines still apply worldviews that predate current understanding of complex system dynamics and of how human evolutionary history has developed with, and helped shape, natural phenomena (5, 6). Their intellectual legacies need to be updated and integrated within an organized scientific effort spanning a range of disciplines that are currently not in effective communication. This would provide essential input to the sustainability discourse.

Resource Imbalances

Little demographic literature addresses the role of resource imbalances as a putative root cause for

> ### KEY TERM
>
> The term **consilience** was coined by William Whewell, in *The Philosophy of the Inductive Sciences*, 1840. The converse of consilience is reductionism. The reductionist modern view is that each branch of knowledge studies a subset of reality that depends on factors studied in other branches. The term was revived in 1998 by humanist biologist E. O. Wilson as an attempt to bridge the culture gap between the sciences and the humanities. Wilson's assertion was that the sciences, humanities, and arts have a common goal: to give a purpose to understanding the details, to lend to all inquirers "a conviction, far deeper than a mere working proposition, that the world is orderly and can be explained by a small number of natural laws."

some of the changes observed in fertility and regional life expectancies (7, 8). The notion of human carrying capacity (9) is generally dismissed as irrelevant (10–12), as if humans, uniquely among species, have transcended environmental dependency. It is true that humans, through cultural developments such as agriculture, trade, and fossil-fuel combustion, have increased the carrying capacity of local environments, at least in the short to medium term. We may yet raise those limits further, or we may now be seeing early evidence of having recently exceeded the global carrying capacity, new technologies notwithstanding. We do not yet know which. Meanwhile, demographers display little awareness of the likely impacts of global environmental changes on future changes in human population size (13). The recent decline in global population growth rate has been generally welcomed by demographers (though noting the attendant problems of population aging and increased dependency ratios), suggesting that, for some, the issue of sustainability is recognized. The world view of many demographers still inclines toward that of many economists in assuming a setting that is free of the constraints of the human carrying capacity of the biosphere.

Market Forces

The role of market forces is central to modern economics, and the turnover of goods and services is considered an indicator of progress. Instead of recognizing that the human economy is a dependent subset of the biosphere, many economists still assume that economic growth and liberalization, with wealth creation, is the key to affording adequate environmental management. Environmental quality is believed to be most effectively achieved through market forces, even as social and environmental costs are "externalized." This view also assumes that environmental change is generally incremental, thereby overlooking the time-lagged, threshold, and irre-

KEY TERM

In systems, a threshold effect is a harmful or even fatal reaction to exceeding the tolerance limit. The consequences, through slow accumulation, can lead to a sudden change. An example of an ecosystemic threshold is when overfishing leads to a sudden collapse of fishery stock. A genocide is an example of an extreme social threshold effect.

versible effects that characterize many changes in human and ecological systems (14).

The growing interdisciplinary domain of environmental and ecological economics appreciates the significance of the Earth system's functioning for human well-being, and, therefore, the need to sustain its capacity to support economic development (15, 16). The economics of complex system dynamics and its implications for sustainability have also been addressed (17). Indeed, ecological economics treats environmental sustainability and human carrying capacity as central premises for economic development (18).

Ecosystems and Human Society

Ecologists understand the structure, functioning, and interdependencies of populations and ecosystems and, increasingly, appreciate the interplay of the natural world with human systems. However, various conceptual and theoretical frameworks in ecology still disregard the connection of ecosystems to the human species. More integrated views from landscape ecology and systems approaches, and the greater appreciation of complex systems, critical thresholds, and the possibilities of state changes, are attracting attention (19, 20).

Over the past decade, the fledgling field of "ecosystem health" has been fostered in an interdisciplinary fashion (21, 22). There is increasing recognition that humans are themselves a major force in ecosystem development and evolution.

Integrative approaches to coevolving social-ecological systems have emerged (23, 24). The Millennium Ecosystem Assessment Project, funded by several international environmental-biological conventions and other international agencies, has brought together many scientists to address interdisciplinary questions relating to the current and future conditions of the world's ecosystems and the consequences for human societies (25).

Risk of Disease

During the recent development of epidemiology as a modern discipline, populations have been increasingly viewed as aggregations of individuals exercising free choices. Accordingly, contemporary epidemiology has focused on quantifying the contribution of specific individual-level factors to disease risk. However, the resurgence of infectious disease, including particularly HIV/AIDS, has underscored the importance of population-level phenomena, including social conditions, cultural practices, and technological choices. Similarly, dramatic changes in health and life expectancy in the countries of central and eastern Europe and the former Soviet bloc, following the collapse of communism, highlight the fundamental importance of social, economic, and political conditions to population health (26, 27). Meanwhile, there is nascent recognition that climate change and other global environmental changes pose risks to human health, both now and, more so, in the future (28).

Responding to the Crisis

Addressing sustainability is more than an academic exercise. It is a vital response to a rapidly evolving crisis and should be at the top of our research agendas. The forces that oppose social change for sustainability, whether from indifference, incomprehension, or self-interest, are powerful. But neither individual scientists nor isolated scientific disciplines will suffice to change understanding and policy. Science itself needs to be fully engaged in this challenge (29). The "science of human-environment interactions" (30) and sustainability science have emerged over the past decade (31). A combination of inter- and transdisciplinary approaches to sustainability, unconstrained by traditional disciplinary domains and concepts, must be encouraged.

Such approaches may prove difficult to achieve within conventional university departments. Purpose-built interdisciplinary centers

> ## KEY TERM
>
> **Sustainability science** is an emerging science that seeks to understand the fundamental character of interactions between nature and society. It encompasses the interaction of global processes with the ecological and social characteristics of particular places and sectors. It addresses such issues as the behavior of complex self-organizing systems as well as the responses, some irreversible, of the nature-society system to multiple and interacting stresses.

will therefore be needed. Other support will come from interdisciplinary societies (e.g., International Association for the Study of Common Property), research institutes (e.g., Santa Fe Institute; Beijer Institute, Stockholm; National Center for Ecological Analysis and Synthesis, Santa Barbara; International Institute for Applied Systems Analysis, Vienna), and research networks [e.g., Sustainability Science network on vulnerability, Resilience Alliance, International Council for Science (ICSU) initiative on sustainability, International Geosphere Biosphere Program, and International Human Dimensions Program on Global Environmental Change]. Achieving a sufficiently intensive interdisciplinary collaboration, on a large enough canvas to meet the needs of sustainability, remains the central challenge.

References

1. P. H. Raven, *Science* 297, 954 (2002).
2. C. D. Butler, *Glob. Change Hum. Health* 1, 156 (2000).
3. E. Ostrom *et al.*, *The Drama of the Commons* (National Academy Press, Washington, DC, 2002).

4. L. White, *Science* 155, 1203 (1967).

5. C. L. Redman, *Human Impact on Ancient Environments* (Univ. of Arizona Press, Tucson, AZ, 1999).

6. R. J. McIntosh et al., *The Way the Wind Blows* (Columbia Univ. Press, New York, 2000).

7. R. A. Easterlin, *Am. Econ. Rev.* 61, 399 (1971).

8. V. D. Abernethy, *Population Politics: The Choices That Shape Our Future* (Plenum, New York, 1993).

9. G. Daily, P. R. Ehrlich, *Bioscience* 42, 761 (1992).

10. C. André, J. P. Platteau, *J. Econ. Behav. Org.* 34, 1 (1998).

11. C. D. Butler, *Ecosyst. Health* 6, 171 (2000).

12. J. E. Cohen, *Science* 269, 341 (1995).

13. P. Demeny, *Popul. Dev. Rev.* 14, 213 (1988).

14. B. Walker et al., *Ecol. Sci.* 9, 5 (1995).

15. K. Arrow et al., *Science* 268, 520 (1995).

16. K. G. Mäler, *Eur. Econ. Rev.* 44, 645 (2000).

17. W. B. Arthur, *Science* 284, 107 (1999).

18. R. Costanza et al., *The Development of Ecological Economics* (Elgar, London, 1997).

19. S. A. Levin, *Fragile Dominion: Complexity and the Commons* (Perseus Books, Reading, MA, 1999).

20. M. Scheffer et al., *Nature* 413, 591 (2001).

21. D. Rapport, R. Costanza, A. J. McMichael, *Trends Ecol. Evol.* 13, 397 (1998).

22. A. Aguirre et al., *Conservation Medicine: Ecological Health in Practice* (Oxford Univ. Press, New York, 2002).

23. L. H. Gunderson, C. S. Holling, *Panarchy: Understanding Transformations in Human and Natural Systems* (Island, Washington, DC, 2002).

24. F. Berkes et al., *Navigating Social-Ecological Systems: Building Resilience for Complexity and Change* (Cambridge Univ. Press, Cambridge, 2003).

25. Millennium Assessment Web site: www.millenniumassessment.org.

26. A. J. McMichael, *Am. J. Epidemiol.* 149, 887 (1999).

27. L. Berkman, I. Kawachi, *Social Epidemiology* (Oxford Univ. Press, Oxford, 2001).

28. A. J. McMichael, R. Beaglehole, *Lancet* 356, 495 (2000).

29. J. Lubchenco, *Science* 279, 491 (1998).

30. P. C. Stern, *Science* 260, 1897 (1993).

31. National Research Council, *Our Common Journey: A Transition Toward Sustainability* (National Academy Press, Washington, DC, 1999).

Web Resources

www.sciencemag.org/cgi/content/full/302/5652/1919/DC1

The Burden
of Chronic Disease

C. G. NICHOLAS MASCIE-TAYLOR and ENAMUL KARIM

The shift from acute infectious and deficiency diseases to chronic noncom-
municable diseases is not a simple
transition but a complex and dynamic epidemiological process, with some diseases disappearing
and others appearing or reemerging. The
unabated pandemic of childhood and adulthood
obesity and concomitant comorbidities are affecting both rich and poor nations. Infectious diseases remain an important public health problem,
particularly in developing countries. More attention should be given to the high burden of disease associated with soil-transmitted helminths
and schistosomiasis, which until recently was not
considered a priority even though regular drug
treatment is obtainable at relatively little cost. In
developing countries, the pressing requirement
is to provide an accessible and good-quality

health-care system; in industrialized countries a
major need is for greater public health education
and the promotion of healthy lifestyles.

The burden of disease and injury attributable to undernutrition, poor water supply, poor
sanitation, and inadequate personal and domestic hygiene accounts for almost 23 percent of the
disability-adjusted life years (DALY) from all
causes worldwide and for 26 percent of DALY
in developing regions (1). International initiatives are targeted primarily at conditions that
cause higher mortality (such as AIDS, tuberculosis, malaria, and vaccine-preventable diseases),
but there is also a need to focus attention on controlling conditions such as soil-transmitted
helminths and schistosomiasis that lead to considerable morbidity.

Until recently, it was thought that human populations were experiencing a simple epidemiological transition. This idea, first put forward by
Omran (2), envisaged three stages—"the age of
pestilence and famine," "the age of receding

This article first appeared in *Science* (12 December 2003: Vol. 302, no. 5652). It has been revised and updated for this edition.

Schistosomiasis is a disease caused by the eggs of parasitic worms. Infection occurs when the human host comes in contact with contaminated fresh water in which certain types of snails that carry schistosomes are living. Eggs travel to the liver or pass into the intestine or bladder and are sometimes found in the brain or spinal cord, where they can cause seizures, paralysis, or spinal cord inflammation. For people who are repeatedly infected for many years, the parasite can damage the liver, intestines, lungs, and bladder. Although schistosomiasis is not found in the United States, 200 million people are infected worldwide.

vector-borne diseases, dengue, dengue hemorrhagic fever, yellow fever, plague, malaria, leishmaniasis, rodent-borne viruses, and arboviruses are persisting, and sometimes reemerging, with serious threats to human health. For example, malaria, which is the foremost vector-borne disease worldwide, continues to worsen in many areas; there are now an estimated 300 million to 500 million cases of malaria worldwide each year with 2 million to 4 million deaths. Since 1975, dengue fever has surfaced in huge outbreaks in more than 100 countries, with as many as 100 million cases each year. These increases reflect societal changes arising from population growth, ecological and environmental changes, and especially suburbanization, together with widespread and frequent air travel. The prevalence of obesity, with its known increased risk of developing chronic ailments, some forms of cancer, type 2 diabetes, and cardiovascular disease, is increasing in most countries. It is estimated that more than 1 billion adults worldwide are overweight and that 300 million are clinically obese. In the United Kingdom, obesity has tripled in the past 20 years, and about two-thirds of adults are overweight.

In the United States, 20 states have obesity prevalence rates of 15 to 19 percent, 29 have rates of 20 to 24 percent, and one has a reported rate of more than 25 percent. Overweight and obesity are not confined to adults, and there is evidence of an increase in the prevalence of childhood overweight and obesity in both developed and developing countries.

pandemics," and "the age of degenerative and man-made diseases"—and assumed that as infectious diseases are eliminated, chronic diseases will increase as the population ages. However, chronic diseases are emerging as a major epidemic in many nonindustrialized countries because of their association with overweight and obesity. In addition, the upsurge of infectious diseases and the emergence of new ones also casts doubt on this simple, unidirectional epidemiological process.

Emerging and Reemerging Disease

A recent review (3) suggested that 175 human pathogens (12 percent of those known) were emerging or reemerging and that 37 pathogens have been recognized since 1973, including rotavirus, Ebola virus, HIV-1 and HIV-2, and most recently, Nipah virus. Among the infectious

Vector-borne disease refers to a category of disease that is transmitted by an insect or any living carrier. The ranges of certain vector-borne diseases, such as malaria, appear to be growing as a result of climate change.

Helminths and Morbidity

Infection by soil-transmitted helminths has been increasingly recognized as an important public health problem, particularly in developing countries. Parasitic infection accounts for an estimated 22.1 million life years lost to hookworm (either *Necator americanus* or *Ancylostoma duodenale*), 10.5 million life years lost to roundworm (*Ascaris lumbricoides*), 6.4 million life years lost to whipworm (*Trichuris trichiura*), and 4.5 million life years lost to schistosomiasis (4). These figures take into account the range of morbidity associated with these infections and with hookworm-induced anemia. The total for all three soil-transmitted infections and schistosomiasis is 43.5 million life years lost, which is second only to tuberculosis (46.5 million) and well ahead of malaria (34.5 million) and measles (34.1 million).

Worm transmission is enhanced by poor socioeconomic conditions, deficiencies in sanitary facilities, improper disposal of human feces, insufficient supplies of potable water, poor personal hygiene, substandard housing, and lack of education [hence intestinal parasitism's label as the "disease of poverty" (5)]. About 25 percent of the world's population is infected with roundworm, 20 percent with hookworm, 17 percent with whipworm, and 3 to 4 percent with schistosomes (Table 6)—overall ~2 billion people worldwide (a third of the world's population) are affected with one or more of these soil-transmitted infections and schistosomiasis (6).

The estimated worldwide percentages of the three intestinal parasites have remained nearly constant over the past 50 years (7), but there have

been some successes. For example, in Japan, intestinal helminth infections were virtually wiped out in a 15- to 20-year period after World War II through an integrated program of education, improved sanitation and water supply, and drug treatment (8).

What has changed over the past 25 years, however, is the recognition that these soil-transmitted helminth infections and schistosomiasis have serious health consequences ranging from reversible growth faltering, permanent growth retardation, and clinically overt symptoms (e.g., nausea, diarrhea, dysentery, and fever) to acute complications (e.g., intestinal obstruction; rectal prolapse; granulomas in the mucosa of the urogenital system, intestine, and liver; cancer of the bladder; hepatomegaly; and ascites). These parasitic infections are associated with malnutrition and impaired growth and development (caused by decreased appetite, nutrient loss, malabsorption, and decreased nutrient utilization), iron deficiency anemia (the bloodsucking activities of hookworm leads to blood loss of between 0.03 and 0.15 ml per day per worm), decreased physical fitness and work capacity, and impaired cognitive function (9–11).

TABLE 6. Estimated global prevalences and associated morbidity and mortality due to soil-transmitted helminths and schistosomes.

Parasite	Prevalence of infection (cases, millions)	Mortality (deaths, thousands)	Morbidity (cases, millions)
Ascaris lumbricoides	1450	60	350
Trichuris trichiura	1050	10	220
Hookworms	1300	65	150
Schistosomes	200	20	20

The formal recognition of the health consequences of worm infestation came as recently as May 2001 when the 54th World Health Assembly passed a resolution affirming that the control of schistosomiasis and soil-transmitted helminthiasis should be considered a public health priority. The challenge ahead is converting words into deeds through a global helminth control program (12).

Global Helminth Control

Horton (13) calculated that ~500 million children will have to be treated regularly for ascariasis for the next 25 years for the absolute numbers to stay the same. However, for a reduction to occur, at least 1 billion will need regular treatment. This figure is for ascariasis alone; to treat all soil-transmitted infections and schistosomiasis would require doubling this number. Furthermore, there is evidence of drug resistance in animals in which intensive helminth control measures have been used, and some concerns have been expressed that the same might happen with humans, although extrapolation from animals raised for food production to humans must take into account genetic and epidemiological differences (14). There are already indications that schistosomiasis cure rates (using praziquantal) are worse than those a decade ago (15), and mebendazole and pyrantel may be less effective against hookworm now than in the past (16, 17). No alternatives exist to praziquantal, but using combinations of drugs or cycling their use may reduce drug resistance.

At first sight, the drug and infrastructure cost of global helminth control appears enormous. For example, the cost of drugs alone in treating 2 billion people annually will be about US$100 million. This sum, although large, has to be seen within the context of worldwide health expenditure per capita, which ranges between about US$12 and US$2,769 (18), whereas the cost of a single-dose antihelmintic treatment is only about US$.03 per annum (and about US$.20 to US$.30 for praziquantal), excluding delivery costs (19).

The current laudable goal of helminth control

RANDOM SAMPLES

Science, Vol 287, Issue 5460, 1917,
17 March 2000

GUINEA WORM BANISHED FROM INDIA

India has finally conquered Guinea worm, making it the second disease after smallpox to be fully eradicated from the country. The disease, which affects mainly the rural poor, is now confined to just 12 African nations, according to an announcement by the World Health Organization (WHO).

Guinea worm, transmitted through contaminated drinking water, is known to researchers as the giant parasitic roundworm *Dracunculus medinensis*, a spaghetti-thin wriggler that can grow up to a meter long. The worm spawns in water, where its eggs are taken up and nurtured by Cyclops, an aquatic insect.

Fifty years ago, 25 million Indians suffered from dracunculiasis, a painful, untreatable condition that can cause crippling infections. But a worm-eradication campaign reduced the number of cases to 40,000 by the early 1980s. And Subhash Salunke, a WHO communicable diseases consultant based in New Delhi, says that "if all goes well in the next 4 to 5 years," the world will be Guinea worm–free by 2010.

is very different from the earlier, but disastrous, attempts at hookworm and malaria eradication. So far only smallpox has been eradicated (1980), but the World Health Organization is also committed to eradicating poliomyelitis (by 2005) and dracunculiasis (guinea worm), and a global lymphatic filariasis campaign has also commenced (20). The dracunculiasis eradication campaign (21) began in 1980, and the incidence fell from an estimated 3.2 million cases in 1986 to 64,000 cases in 2001. More than 150 countries and territories have been certified free of parasite transmission, and the eradication goal is in sight. However, programs are being disrupted, particularly in countries affected by civil conflict, such as Sudan, where ~78 percent of the world's cases were reported in 2001.

Global Health Trajectories and Solutions

The health and disease patterns of societies and countries evolve as a result of socioeconomic, demographic, technological, cultural, environmental, and biological changes. Wars and civil conflict continue to disrupt the human host-agent-environment equilibrium and elevate disease burden, and injuries caused by accidents and violence are also increasing.

So is "Health for All in the 21st Century" (22) any closer? The industrialized countries appear to be exchanging one enemy for another: Having in the main brought infectious diseases under control, the unabated increase in obesity in childhood and adulthood and its concomitant comorbidities are likely to result in massive social and economic burdens. Only with dramatic changes in lifestyle—decreases in food portion sizes, energy density of the diet, and fat intake, and increases in fruit and vegetable consumption and physical activity—can obesity and the metabolic syndrome epidemic be brought under control. Successful strategies involve governments and local communities working together to initiate programs in schools, the workplace, and communities (23) and should involve food producers, the food-processing industry, and consumer associations.

Improving the health status of poor populations requires a twin approach. Not only are infectious diseases still common, but chronic diseases, including tobacco-related diseases, are on the rise. Many of the poorer countries lack accessible, affordable, and high-quality health-care systems.

Strengthening national health policies, managing and mobilizing resources, training personnel, and providing service delivery are key goals. Developing public health strategies at national and global levels and financing and organizing them will continue to present enormous challenges (24).

References

1. C. J. L. Murray, A. D. Lopez, *The Global Burden of Disease* (Harvard University Press, Cambridge, MA, 1996).

2. A. R. Omran, *Milbank Mem. Fund Q.* 49, 509 (1971).

3. D. J. Gubler, *Emerg. Infect. Dis.* 4, 442 (1998).

4. M.-S. Chan, *Parasitol. Today* 13, 438 (1997).

5. E. Cooper, *Trans. R. Soc. Trop. Med. Hyg.* 85, 168 (1991).

6. A. Montressor, D. W. T. Crompton, T. W. Gyorkos, L. Savioli, *Helminth Control in School-Age Children* (World Health Organization, Geneva, 2002).

7. N. R. Stoll, *J. Parasitol.* 33, 1 (1947).

8. M. Yokogawa, in *Ascariasis and its Public Health Significance*, D. W. T. Crompton, Ed. (Taylor and Francis, London, 1985), pp. 265–277.

9. D. W. T. Crompton, *Parasitology* 121, S39 (2000).

10. L. S. Stephenson, M. C. Latham, E. A. Ottesen, *Parasitology* 121, S23 (2000).

11. P. O'Lorcain, C. V. Holland, *Parasitology* 121, S51 (2000).

12. L. Savioli, *Trans. R. Soc. Trop. Med. Hyg.* 86, 353 (2002).

13. J. Horton, *Trends Parasitol.* 19, 405 (2003).

14. S. Geerts, B. Gryseels, *Trop. Med. Int. Health* 6, 915 (2001).

15. D. Cioli, *Curr. Opin. Infect. Dis.* 13, 659 (2000).

16. M. Sacko et al., *Trans. R. Soc. Trop. Med. Hyg.* 93, 195 (1999).

17. J. A. Reynoldson et al., *Acta Trop.* 68, 301 (1997).

18. Human Development Reports, United Nations Development Programme, www.undp.org/hdr2003/indicator/indic_59_1_1.html.

19. H. Guyatt, in *The Geohelminths: Ascaris, Trichuris, and Hookworm*, C. V. Holland, M. W. Kennedy, Eds. (Kluwer Academic Publishers, Dordrecht, Netherlands, 2002), pp. 75–87.

20. The Global Alliance to Eliminate Lymphatic Filariasis, www.filariasis.org/index.pl?id=2586.

21. Centers for Disease Control and Prevention, *Morb. Mortal. Wkly. Rep.* 52, 881 (2003), www.cdc.gov/mmwr/preview/mmwrhtml/mm5237a1.htm.

22. World Health Organization, Health for All in the 21st Century, www.who.int/archives/hfa.

23. Rhode Island Department of Health, Obesity Control Program, www.healthri.org/disease/obesity/Home.htm.

24. R. D. Smith, *Bull. WHO* 81, 475 (2003).

Web Resources

www.sciencemag.org/cgi/content/full/302/5652/1921/DC1

The Challenge of Long-Term Climate Change

K. HASSELMANN, M. LATIF, G. HOOSS, C. AZAR,

O. EDENHOFER, C. C. JAEGER, O. M. JOHANNESSEN,

C. KEMFERT, M. WELP, and A. WOKAUN

Climate policy needs to address the multidecadal to centennial time scale of climate change. Although the realization of short-term targets is an important first step, to be effective climate policies need to be conceived as long-term programs that will achieve a gradual transition to an essentially emission-free economy on the time scale of a century. This requires a considerably broader spectrum of policy measures than the primarily market-based instruments invoked for shorter-term mitigation policies. A successful climate policy must consist of a dual approach that focuses on both short-term targets and long-term goals.

There is widespread consensus in the climate research community that human activities are changing the climate through the release of greenhouse gases, particularly CO_2, into the atmosphere (1, 2). Because of the considerable inertia of the climate system—caused by the long residence times of many greenhouse gases in the atmosphere, the large heat capacity of the oceans, and the long memory of other components of the climate system, such as ice sheets and the biosphere—human modifications of the climate system through greenhouse gas emissions are likely to persist for many centuries in the absence of appropriate mitigation measures (2).

A common response to the uncertain risks of future climate change is to develop climate policy as a sequence of small steps. The Kyoto Protocol commits the signatory countries to a nominal reduction of greenhouse gas emissions by 5 percent between 2008 and 2012, relative to 1990. The protocol is a historic first step toward reversing the trend of continually increasing greenhouse gas emissions and will provide valuable experience in the application of various mit-

This article first appeared in *Science* (12 December 2003: Vol. 302, no. 5652). It has been revised and updated for this edition.

igation instruments such as tradable emission permits. However, a nominal emission reduction of only 5 percent by a subset of the world's nations will have a negligible impact on future global warming. To avoid major long-term climate change, average per capita greenhouse gas emissions must be reduced to a small fraction of the present levels of developed countries on the time scale of a century (2). Such reductions cannot be achieved by simply extrapolating short-term policies but require a broader spectrum of instruments.

Most investigations (2–4) and public attention have focused on the projected climate change in this century. A potentially far more serious problem, however, is the global warming anticipated in subsequent centuries if greenhouse gas emissions continue to increase unabated (Figure 27, left panels) (5–7). The projected temperature and sea level changes for the next millennium greatly exceed those in the next hundred years (Figure 27, blue boxes). If all estimated fossil fuel resources are burnt, CO_2 concentrations between 1,200 parts per million (ppm) (scenario C in Figure 27) and 4,000 ppm (scenario E in Figure 27) are predicted in the second half of this millennium, leading to temperature increases of 4°C to 9°C and a sea level rise of 3 to 8 m. Predictions of this magnitude are beyond the calibration ranges of climate models and must therefore be treated with caution (8). However, the predicted climate change clearly far exceeds the natural climate variability (~1°C to 2°C) experienced in the past 10,000 years. Even if emissions are frozen at present levels, the accumulated emissions over several centuries still yield climate change on the order of the lower business-as-usual (BAU) scenario C.

Major climate change can be avoided in the long term only by reducing global emissions to a small fraction of present levels within one or two centuries. As an example, we have computed optimal CO_2 emissions paths that minimize the time-integrated sum of climate damage and mitigation costs, using an integrated assessment model consisting of a nonlinear impulse response climate model (7) coupled to an elementary economic model (9) (Figure 27, right panels).

Cost-benefit analyses depend on many controversial assumptions, such as the role of economic inertia (included in case a, ignored in case b), the impact of declining costs for new technologies, and the discount factors applied to future climate change mitigation and adaptation costs (10–15). However, the resultant long-term climate change is insensitive to the details of the optimal emission path (compare curves a and b), provided the emissions are sufficiently reduced. Because of the long residence time of CO_2 in the atmosphere (>100 years), the climatic response

KEY TERMS

Integrated assessment (IA) of global climate change is the analysis of climate change from the cause, such as greenhouse gas emissions, through effects, such as the need to increase air conditioning because of temperature changes. The analysis emphasizes feedbacks. It integrates the assessment of human behavior with that of biophysical factors. IA is usually, but not always, implemented as a computer model.

An optimal emissions path is the balance of policy measures that minimizes the sum of the damages caused by climate change and the costs of reducing emissions. Integrated assessment models are used to help design policies aimed at achieving optimal emissions paths.

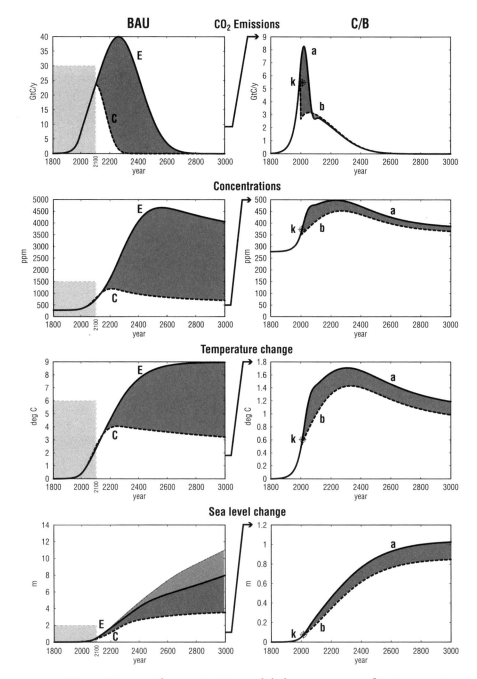

FIGURE 27. CO$_2$ emissions and concentrations, global mean near surface temperature, and global mean sea level for business-as-usual (BAU) emission scenarios (left) and optimized cost/benefit (C/B) trajectories (right; note change of scale). The BAU scenarios assume that all fossil fuel resources, ranging from 4,000 gigatons of carbon (GtC) (conventional resources, C) to 15,000 GtC (conventional plus exotic resources, E), are used. The sea level rise represents the sum of thermal expansion of the warming ocean, the melting of smaller inland glaciers, and the slow melting of the Greenland Ice Sheet. Including other green-house gases could increase the peak values by about 10 to 20 percent. The cost/benefit solutions include (a) or ignore (b) economic inertia. Pronounced differences between these cases in the short term have little impact on long-term climate. The impact of the Kyoto period (k) is not discernible on these multicentennial time scales.

is governed by the cumulative CO_2 emissions rather than by the detailed path.

The impact of the Kyoto agreement (k in Figure 27, right panels) is hardly discernible on the millennial time scale, suggesting that the Kyoto debate should focus on the long-term implications of the protocol rather than on its short-term effectiveness. The Kyoto targets may not be met by some countries and may be exceeded by others. Important in either case is that the Kyoto policy is accompanied by measures that ensure continuing reductions in subsequent decades.

Because of the 10-year horizon of the Kyoto Protocol, climate policy has tended to focus on promoting mitigation technologies that are currently most cost-effective, such as wind energy, biomass fuels, fuel switching from coal and oil to gas, and improved energy efficiency in transportation, buildings, and industry. In the short to medium term, the combined mitigation potential of these technologies is substantial. It has been estimated that, if fully implemented, together they could halve global greenhouse gas emissions relative to the business as usual (BAU) level within two decades (4). The market-based instruments (such as tradable emission permits and tax incentives; see the chapters "Energy Resources and Global Development" and "The Struggle to Govern the Commons"), used to meet the more modest 5 percent Kyoto reduction targets will accelerate the penetration of these technologies into the marketplace but will be inadequate to realize the full potential of these technologies.

Yet, even if forcefully implemented, currently available low-cost technologies have a limited capacity for substantial global emission reduction and will not be able to counter the rising emissions projected for the long term. Future emissions will be driven mainly by the expanding populations of the developing world, which strive to achieve the same living standards as the industrial countries. An emissions reduction of 50 percent applied to a projected BAU increase in this century by a factor of four (2–4) still leads to a doubling of emissions, far from the long-term target of near-zero emissions.

Furthermore, the mitigation costs for today's technologies are estimated to rise rapidly if per capita emissions are reduced by more than half (4). Thus, although the Kyoto Protocol will boost technologies that are cost-effective in the short term, further emission reductions in the post-Kyoto period could be limited by prohibitive costs. Without affordable new technologies capable of higher global emission reductions, stricter emission reduction targets will be considered impossible to meet and will not be adopted. Although no such technology is yet economically

KEY TERM

Carbon sequestration: The long-term storage of carbon, either in living systems, such as rainforests, which absorb carbon as part of the evapo-respiratory process, or through the physical "scrubbing" of carbon from emissions and storage in geologic cavities or at the bottom of the oceans are among other options that have been proposed.

competitive, there exist many promising candidates (16, 17), ranging from solar thermal or photovoltaic energy—in combination with hydrogen technology—to **carbon sequestration** in geological formations or the ocean (18–21), advanced nuclear fission, and nuclear fusion (4, 16, 17). Which technology, or mix of technologies, will ultimately prove most cost-effective cannot be predicted. We will need to accept these uncertainties and support a number of competing technologies in order to have available several commercially viable alternatives when the large-scale transition to low-emission technologies becomes more urgent.

Although short-term climate policy can be

formulated in terms of emission targets and implemented with instruments that force emit-

> Without affordable new technologies capable of higher global emission reductions, stricter emission reduction targets will be considered impossible to meet and will not be adopted. Although no such technology is yet economically competitive, there exist many promising candidates.

ters to bear the costs incurred by climate change (the "polluter-pays principle"), long-term climate policy will require a broader spectrum of measures extending well beyond the traditional horizon of government policies or business investment decisions. The entry of new technologies into the marketplace depends on multiple incentives and feedbacks, including private investments, government investments in infrastructure and subsidies for pilot plants, protected niche markets, and changes in consumer preferences and lifestyles (22–26). Climate is a public good that demands communal action for its protection, including the involvement of citizens and institutions such as the media that shape long-term public attitudes. Self-interest alone will not motivate businesses and the public to change established practices and behavioral patterns. The goal of long-term climate policy must be to influence business investments,

research, education, and public perceptions such that stringent emission-reduction targets — although not attainable today — become acceptable at a later time.

Although major changes are necessary, the long time scales of the climate system allow a gradual transition (27, 28). Estimated costs to halve global emissions range from ~1 to 3 percent of gross domestic product (GDP) (4), similar to the annual GDP growth rate in many countries. Thus, implementing an effective climate policy over a time period of, say, 50 years would delay economic growth by only about a year over the same period (29). This appears to be an acceptable price for avoiding the risks of climate change. However, because the global political-economic system exhibits considerable inertia, a transition to a sustainable climate can be achieved without major socioeconomic dislocations only if the introduction of appropriate measures that address the long-term mitigation goals is not delayed.

Science can assist the development of long-term climate policies by providing detailed analyses of the technological options and their implications for national economies and global development. The Intergovernmental Panel on Climate Change (IPCC) has played a pivotal role in the climate debate by presenting authoritative reviews of the state of science and of climate change impact, mitigation, and policy. Similar expertise should be made available to climate negotiators in the form of timely analyses of the implications of alternative climate policy regimes for the individual signatories of the original United Nations Framework Convention on Climate Change, adopted at the Earth Summit in Rio de Janeiro in 1992.

Although binding long-term commitments cannot be expected from governments, declarations of long-term policy goals and visible actions to achieve these goals are essential for the investment plans of businesses, particularly for energy technologies characterized by long capital lifetimes. A long-term perspective is equally important for the public, who must understand and support the policies.

Binding commitments to meet short-term emission-reduction targets must therefore go hand in hand with clearly defined strategies to achieve substantially more stringent reductions in the longer term.

References and Notes

1. T. P. Barnett *et al.*, *Bull. Am. Meteor. Soc.* 80, 2631 (1999).
2. Intergovernmental Panel on Climate Change (IPCC), *Climate Change 2001, Working Group 1: The Scientific Basis*, J. T. Houghton *et al.*, Eds. (Cambridge University Press, 2001).

SCIENCE IN THE NEWS

Science, Vol. 309, no. 5731, 100, 1 July 2005

HOW HOT WILL THE GREENHOUSE WORLD BE?
Richard A. Kerr

Scientists know that the world has warmed lately, and they believe humankind is behind most of that warming. But how far might we push the planet in coming decades and centuries? That depends on just how sensitively the climate system—air, oceans, ice, land, and life—responds to the greenhouse gases we're pumping into the atmosphere. For a quarter-century, expert opinion was vague about climate sensitivity. Experts allowed that climate might be quite touchy, warming sharply when shoved by one climate driver or another, such as the carbon dioxide from fossil fuel burning, volcanic debris, or dimming of the sun. On the other hand, the same experts conceded that climate might be relatively unresponsive, warming only modestly despite a hard push toward the warm side.

Climate Drivers: Carbon emissions caused by human activities such as driving cars may be having a measurable effect on the global climate.
CREDIT: PAIWEI WEI/ISTOCKPHOTO.COM

The problem with climate sensitivity is that you can't just go out and directly measure it. Sooner or later a climate model must enter the picture. Every model has its own sensitivity, but each is subject to all the uncertainties inherent in building a hugely simplified facsimile of the real-world climate system. As a result, climate scientists have long quoted the same vague range for sensitivity: A doubling of the greenhouse gas carbon dioxide, which is expected to occur this century, would eventually warm the world between a modest 1.5°C and a whopping 4.5°C. This range—based on just two early climate models—first appeared in 1979 and has been quoted by every major climate assessment since.

Researchers are finally beginning to tighten up the range of possible sensitivities, at least at one end. For one, the sensitivities of the available models (5 percent to 95 percent confidence range) are now falling within the canonical range of 1.5°C to 4.5°C; some had gone considerably beyond the high end. And the first try at a new approach—running a single model while varying a number of model parameters such as cloud behavior—has produced a sensitivity range of 2.4°C to 5.4°C with a most probable value of 3.2°C.

3. IPCC, *Climate Change 2001, Working Group 2: Impacts, Adaptation, and Vulnerability*, J. J. McCarthy *et al.*, Eds. (Cambridge University Press, 2001).

4. IPCC, *Climate Change 2001, Working Group 3: Mitigation*, B. Metz *et al.*, Eds. (Cambridge University Press, 2001).

5. W. R. Cline, *The Economics of Global Warming* (Institute for International Economics, Washington, DC, 1992).

6. T. J. Crowley, K.-Y. Kim, *Geophys. Res. Lett.* 22, 933 (1995).

7. G. R. Hooss *et al.*, *Clim. Dyn.* 18, 189 (2001).

8. The uncertainties of climate predictions are estimated to be ~50 percent, excluding instabilities of the climate system that could yield substantially larger changes, for example, through the collapse of the Gulf Stream and thermohaline ocean circulation, a break-off of the West Antarctic ice sheet, or the release of methane presently frozen in permafrost regions.

9. K. Hasselmann *et al.*, *Clim. Change* 37, 345 (1997).

10. W. D. Nordhaus, *Science* 294, 1283 (2001).

11. P. G. Brown, *Clim. Change* 37, 329 (1997).

12. G. Heal, *Clim. Change* 37, 335 (1997).

13. W. D. Nordhaus, *Clim. Change* 37, 315 (1997).

14. K. Hasselmann, *Clim. Change* 41, 333 (1999).

15. C. Kemfert, *Ecological Economics*, 54, 293 (2005).

16. J. Goldemberg, Ed., *World Energy Assessment (2000): Energy and the Challenge of Sustainability* (United Nations Development Programme, United Nations Department of Economics and Social Affairs, World Energy Council, New York, 2001).

17. M. I. Hoffert *et al.*, *Science* 298, 981 (2002).

18. B. P. Eliasson *et al.*, Eds., *Greenhouse Gas Control Technologies* (Pergamon, Amsterdam, 1999).

19. K. S. Lackner, *Science* 300, 167 (2003).

20. P. G. Brewer *et al.*, *Science* 284, 943 (1999).

21. H. Drange *et al.*, *Geophys. Res. Lett.* 28, 2637 (2001).

22. M. Haduong *et al.*, *Nature* 389, 270 (1997).

23. O. Edenhofer *et al.*, *Ecological Economics*, 54, 277 (2005).

24. Examples are tradable renewable energy permits (30) and long-term policies in tradable emission permits (31, 32).

25. B. A. Sandén, C. Azar, *Energy Policy* 33, 1557 (2005).

26. M. Weber, V. Barth, K.Hasselmann, *Ecological Economics*, 54, 306 (2005).

27. J. Alcamo, E. Kreileman, *Glob. Environ. Change* 6, 305 (1996).

28. B. C. O'Neill, M. Oppenheimer, *Science* 296, 1971 (2002).

29. C. Azar, S. H. Schneider, *Ecol. Econ.* 42, 73 (2002).

30. D. Barry, *Ecol. Econ.* 42, 369 (2002).

31. S. C. Peck, T. J. Teisberg, in *Risk and Uncertainty in Environmental and Resource Economics*, J. Wesseler, H-P Weikard, Eds. (Edward Elgar, United Kingdom, in press), chap. 9.

32. M. Leimbach, *Energy Policy* 31, 1033 (2003).

33. The views expressed in this article evolved from discussions with members and guests of the European Climate Forum (ECF). We acknowledge constructive comments from G. Berz, C. Carraro, B. Eliasson, J. Engelhard, J. Gretz, B. Hare, J.-C. Hourcade, M. Hulme, M. McFarlane, N. Otter, H.-J. Schellnhuber, S. Singer, and S. C. Peck. However, ECF does not endorse specific views expressed by its members, and this article does not represent an ECF consensus view.

Web Resources

www.sciencemag.org/cgi/content/full/302/5652/1923/DC1

Climate Change
The Political Situation

ROBERT T. WATSON

Human-induced climate change is one of the most important environmental issues facing society worldwide. The overwhelming majority of scientific experts and governments acknowledge that there is strong scientific evidence demonstrating that human activities are changing the Earth's climate and that further human-induced climate change is inevitable. Changes in the Earth's climate are projected to adversely affect socioeconomic systems (such as water, agriculture, forestry, and fisheries), terrestrial and aquatic ecological systems, and human health. Developing countries are projected to be most adversely affected, and poor people within them are the most vulnerable. The magnitude and timing of changes in the Earth's climate will depend on the future demand for energy, the way it is produced and used, and changes in land use, which in turn affect emissions of greenhouse gases and aerosol precursors.

The most comprehensive and ambitious attempt to negotiate binding limits on greenhouse gas emissions is contained in the 1997 Kyoto Protocol, an agreement forged in a meeting of more than 160 nations, in which most developed countries agreed to reduce their emissions by 5 to 10 percent relative to the levels emitted in 1990. Although the near-term challenge for most industrialized countries is to achieve their Kyoto targets, the long-term challenge is to meet the objectives of Article 2 of the United Nations Framework Convention on Climate Change (UNFCCC)—that is, stabilization of greenhouse gas concentrations in the atmosphere at levels that would prevent dangerous anthropogenic interference with the climate system, with specific attention being paid to food security, ecological systems, and sustainable economic development. To stabilize the atmospheric

This article first appeared in *Science* (12 December 2003: Vol. 302, no. 5652). It has been revised and updated for this edition.

Science, Vol. 305, no. 5686, 962–963, 13 August 2004

THE CARBON CONUNDRUM
Robert F. Service

Even if the hydrogen economy were technically and economically feasible today, weaning the world off carbon-based fossil fuels would still take decades. During that time, carbon combustion will continue to pour greenhouse gases into the atmosphere—unless scientists find a way to reroute them. Governments and energy companies around the globe have launched numerous large-scale research and demonstration projects to capture and store, or sequester, unwanted carbon dioxide (see table). Although final results are years off, so far the tests appear heartening. "It seems to look more and more promising all the time," says Sally Benson, a hydrogeologist at Lawrence Berkeley National Laboratory in California. "For the first time, I think the technical feasibility has been established."

Fossil fuels account for most of the 6.5 billion tons (gigatons) of carbon—the amount present in 25 gigatons of CO_2—that people around the world vent into the atmosphere every year. And as the amount of the greenhouse gas increases, so does the likelihood of triggering a debilitating change in Earth's climate. Industrialization has already raised atmospheric CO_2 levels from 280 to 370 parts per million, which is likely responsible for a large part of the 0.6°C rise in the average global surface temperature over the past century. As populations explode and economies surge, global energy use is expected to rise by 70 percent by 2020, according to a report last year from the European Commission, much of it to be met by fossil fuels. If projections of future fossil fuel use are correct and nothing is done to change matters, CO_2 emissions will increase by 50 percent by 2020.

To limit the amount of CO_2 pumped into the air, many scientists have argued for capturing a sizable fraction of that CO_2 from electric plants, chemical factories, and the like and piping it deep underground. In June, Ronald Oxburgh, Shell's chief in the United Kingdom, called sequestration essentially the last best hope to combat climate change. "If we don't have sequestration, then I see very little hope for the world," Oxburgh told the British newspaper *The Guardian*.

SOME CO_2 SEQUESTRATION PROJECTS

Project	Location	Tons of CO_2 to be Injected	Source	Status
Sleipner	North Sea	20 million	Gas field	Ongoing
Weyburn	Canada	20 million	Oil field	Completed phase 1
In Salah	Algeria	18 million	Gas field	Starts 2004
Gorgon	Australia	125 million	Saline aquifer	In preparation
Frio	U.S.	3000	Saline aquifer	Pilot phase
RECOPOL	Poland	3000	Coal seams	Ongoing

concentration of carbon dioxide requires that emissions eventually be reduced to only a small fraction—5 to 10 percent—of current emissions.

All major industrialized countries except the United States, the Russian Federation, and Australia have ratified the Kyoto Protocol. The United States and Australia have publicly stated that they will not ratify it, and statements from the Russian Federation are contradictory. Russian ratification is essential for the Kyoto Protocol to enter into force.

The United States has stated that the Kyoto Protocol is flawed policy for four reasons:

1. There are still considerable scientific uncertainties. However, although it is possible that the projected human-induced changes in climate have been overestimated, it is equally possible that they have been underestimated. Hence, scientific uncertainties, as agreed by the governments under Article 3 of the UNFCCC, are no excuse for inaction (the precautionary principle).

2. High compliance costs would hurt the U.S. economy. This is in contrast to the analysis of the Intergovernmental Panel on Climate Change (IPCC), which estimated that the costs of compliance for the United States would be between US\$14 and US\$135 per ton of carbon avoided with international carbon dioxide emissions trading (a 5-cents-per-gallon gasoline tax would be equivalent to US\$20 per ton of carbon). These costs could be further reduced by the use of carbon sinks, by carbon trading with developing countries, and by the reduction of other greenhouse gas emissions.

3. It is not fair, because large developing countries such as India and China are not obligated to reduce their emissions. However, fairness is an equity issue. The parties to the Kyoto Protocol agreed that industrialized countries had an obligation to take the first steps to reduce their greenhouse gas emissions, recognizing that ~80 percent of the total anthropogenic emissions of greenhouse gases have been emitted from industrialized countries (the United States currently emits ~25 percent of global emissions); that per capita emissions in industrialized countries far

> The need to reduce greenhouse gas emissions offers a unique opportunity to modernize energy systems and enhance competitiveness in a globalized world.

exceed those from developing countries; that developing countries do not have the financial, technological, or institutional capability of industrialized countries to address the issue; and that increased use of energy is essential for poverty alleviation and long-term economic growth in developing countries.

4. It will not be effective, because developing countries are not obligated to reduce their emissions. It is true that long-term stabilization of the atmospheric concentration of greenhouse gases cannot be achieved without global reductions, especially because most of the projected growth in greenhouse gas emissions over the next 100 years is from developing countries. Hence, developing countries will have to limit their emissions of greenhouse gases, but industrialized countries should take the lead, as agreed in Kyoto.

Protecting the climate system will require substantial reductions in greenhouse gas emissions. The Kyoto Protocol is recognized to be only the first step on this long journey. However, unless the United States agrees to meaningful reductions in greenhouse gas emissions, it is highly unlikely that major developing countries will agree to limit their emissions or that industrialized countries will agree to additional reductions beyond those already agreed in Kyoto.

One very positive development is that about half of the U.S. states have enacted some climate protection measures, and there are a number of initiatives in the U.S. Congress that would reduce greenhouse gas emissions. Although the McCain-Lieberman Climate Stewardship Act failed to pass in the Senate, 43 senators did vote for it, demonstrating an increasing recognition by members of Congress that there is an urgent need to deal with the climate issue. In addition, more than 40 multinational companies have voluntarily agreed to reduce their emissions of greenhouse gases and to improve the energy efficiency of their products. Several of these companies have already met or exceeded their initial targets and have saved money in doing so.

Technologies exist or can be developed to limit the atmospheric concentration of carbon dioxide to between 450 and 550 parts per million (ppm) in a cost-effective way, but it will take political will, enhanced research and development activities, public-private partnerships, and supporting policies to overcome barriers to the diffusion of these technologies into the marketplace. A number of countries, including the United States, have committed themselves to developing climate-friendly technologies, but the level of investment must be substantially increased. The Kyoto Protocol needs to be ratified, and the United States needs to take meaningful actions to reduce its greenhouse gas concentrations. Governments should then consider setting a long-term target based either on a greenhouse gas stabilization level (between 450 and 550 ppm) or on limits for both the absolute magnitude of global temperature change (less than 2 to 3°C) and the rate of temperature change (less than 0.2°C per decade). A series of intermediate targets can then be developed to involve developing countries in an equitable manner. The need to reduce greenhouse gas emissions offers a unique opportunity to modernize energy systems and enhance competitiveness in a globalized world.

Tales from a
Troubled Marriage
Science and Law in Environmental Policy

OLIVER HOUCK

Early environmental policy depended on science, with mixed results. Newer approaches continue to rely on science to identify problems and solve them, but use other mechanisms to set standards and legal obligations. Given the important role that science continues to play, however, several cautionary tales are in order concerning "scientific management," "good science," the lure of money, and the tension between objectivity and involvement in important issues of our time.

"The scientific debate remains open. Voters believe that there is *no consensus* about global warming within the scientific community. Should the public come to believe that the scientific issues are settled, their views about global warming will change accordingly. Therefore, you

This article first appeared in *Science* (12 December 2003: Vol. 302, no. 5652). It has been revised and updated for this edition.

need to continue to make the lack of scientific certainty the primary issue in the debate . . ." [Frank Luntz, political strategist, 2002 (1)].

This chapter explores the relationship between science and law in environmental policy. The relationship has not been easy, nor has it achieved closure after more than 30 years of marriage. Two alpha partners are still trying to figure out who does what. Both agree on the importance of an environmental policy. The debate is about what it should be based on and how it should be carried out.

Back in the pre-dawn of public environmental statutes, there were private remedies for environmental harms, in tort and nuisance. If someone contaminated your apple orchard, or your child, you could seek damages and even an injunction against the activity. These remedies proved insufficient for at least two reasons. The first is that a civil law response to harm already done is small solace for someone who has lost her livelihood or the health of her child. The second is illustrated by the real-life saga described in *A Civil Action*, involving the contamination of drinking water from, in all probability, industrial waste sites (*2*). Children died, others were rendered vegetables for life, and their parents suffered a grief that is impossible to describe. But their legal case failed, as many others did, over the requirements of proof and causation. Which chemical, of the many toxins in the waste sites, caused these strange infirmities and through exactly what exposure pathways? Which waste sites were responsible: this one, operated by a company with lawyers on tap and a war chest of money available for its defense; or that one, now abandoned, once owned by a corporation long dissolved? Civil law failed because the science could not make the proof.

First-Generation Environmental Law: Science Embraced

Beginning in the 1960s, Congress surmounted these difficulties with new public environmental statutes, each based on standards of performance. The standards would operate by preventing rather than compensating for harm. They would, further, bypass the rigors of causation and proof: Once a standard was set, one had only to see whether or not it was met. The question remained, however: Who would set the standard? The answer seemed apparent. Scientists would, on the basis of scientific analysis. After all, it was the scientists, such as Rachel Carson, Jacques Cousteau, and Yuri Timoshenko, who had sounded the alarm; they were the ones to put out the fire.

The first wave of environmental law, therefore, was science-based environmental policy in action. One of the first was the Water Quality Act of 1965 (*3*), which sought the attainment of water quality criteria. It was soon followed by the National Environmental Policy Act of 1969 (*4*) and the analysis of environmental impact. Then came the Clean Air Act in 1970 (*5*), focused on the attainment of national ambient air quality standards, soon followed by the Resource

Conservation and Recovery Act (RCRA) (waste disposal) (*6*), the Comprehensive Environmental Response, Compensation, and Liability Act (CERCLA) (abandoned waste sites) (*7*), the Toxic Substances Control Act (TOSCA) (chemicals) (*8*), the Federal Insecticide, Fungicide, and Rodenticide Act (FIFRA) (pesticides) (*9*), and the Safe Drinking Water Act (*10*), all with the same premise: Science would tell us what was safe and what was not. Scientists would draw the lines.

It didn't work. None of these laws worked well, and some, after enormous investment, failed utterly (*11*). We began to realize that science, although endlessly fascinating and constantly revelatory, is rarely dispositive. And in the world of environmental policy, that which is not dispositive is dead on arrival. The reason is political: Environmental policy faces a degree of resistance unique in public law. No one who has to comply with environmental law likes it, and many hate it outright. A conventional explanation is money, and that is certainly a factor; it takes more capital to install pollution controls or to raise the causeway on stilts. Environmental law is also intrusive: It involves other people, state bureaucrats for one, in the operation of your oil refinery, pig farm, or real estate portfolio.

Worse, it puts the general public in there too, nosing around, asking questions, taking their complaints to the media. They interfere with your personal life as well: your commuter highway, your garbage, or the way your granddaddy ran cattle and you've always run cattle in your family. These are life choices; often life values. Read George Will (*12*), listen to Rush Limbaugh (*13*). For some, environmental policies seem to threaten their very soul.

Not far from the bottom of all of this resistance is one more element: the embarrassment factor. No one likes to be tagged with the responsibility for poisoning children with lead or destroying the Everglades, and a small industry of euphemisms has sprung up to mask the blame. Strip mining becomes "the removal of overburden," as if the soil, grass, and trees were somehow oppressing the land; dredgers in Louisiana leave "borrow pits," as if they were going to give the soil back someday.

At the top of the 2002 domestic agenda are the "clear skies" and "forest health" initiatives (14), labels that at the least disguise the contents of these programs, if not belie them. This is embarrassment speaking.

The extraordinary degree of resistance to environmental policy brings at least two consequences. First: That which is not nailed down by law is not likely to happen. Second: Even requirements that are nailed down by law, such as the permit requirements of the Clean Water Act or the no-jeopardy standard of the Endangered Species Act (ESA), secure compliance rates of about 50 percent (15). A good rule of thumb is that no environmental law, no matter how stringently written, achieves more than half of what it set out to do.

With this understanding of the special challenges of environmental policy, it is easy to see why science-based approaches fare so poorly. One lesson in this regard can be drawn from the Federal Water Pollution Control Act, aimed at the attainment of water quality standards. Scientists would establish concentration limits for every pollutant, and when waters exceeded these limits, scientists would determine the cause and require abatement.

But concentration limits for what use: swimming, drinking water, or fishing? If for fishing, would the target be catfish or trout, which have widely differing requirements for dissolved oxygen? And for the lower Mississippi River, which basically floats boats, why would one need fish at all? The first question in the standard-setting process, then, depended on identifying goals that were purely political and that, as Congress later found, led inexorably to a race to the bottom: States lowered their standards to attract industry, which then held them hostage under the threat of moving away (16).

The "scientific" part of the act was equally fluid. It involved extrapolating "acceptable" concentration limits from laboratory experiments to natural surroundings; from single pollutants to cocktails of multiple pollutants; and from rapid, observable, lethal effects to long-term, sublethal, and reproductive effects. Then came dilution factors, fate, and dispersion and mixing zones. Conclusions differed by factors of 10, scientist against scientist. When it came next to enforcement, someone had to prove who and what were

> The first question in the standard-setting process, then, depended on identifying goals that were purely political and that, as Congress later found, led inexorably to a race to the bottom: States lowered their standards to attract industry, which then held them hostage under the threat of moving away.

causing the exceedance of the standards. If Lake Pontchartrain turned entropic, was it the cattle farming, the shoe tannery, the local sewage system, or Mother Nature? The higher the stakes, the more contested the science. The problem was not information, it was closure. We had returned to the difficulties of *A Civil Action*. Whether in tort law or public law, the proofs failed.

Environmental statutes addressing toxicity record the problem in a more acute form. In the early 1970s, a number of laws were enacted

SCIENCE IN THE NEWS

Science, Vol. 303, no. 5654, 34, 2 January 2004

UNCERTAIN SCIENCE UNDERLIES NEW MERCURY STANDARDS
Erik Stokstad

The debate has a familiar ring. The Bush Administration, mandated to curb power plant emissions of mercury, unveiled two schemes for reducing the potent neurotoxin. Environmentalists countered that the Environmental Protection Agency's (EPA's) proposals go easy on industry and would do too little too late, "needlessly putting another generation of children at risk of mercury exposure," says Michael Shore of Environmental Defense in New York City.

Rhetoric aside, much of the underlying science is still uncertain. Recent studies do suggest that in some locations cutting emissions can help wildlife—and thus presumably human health— within years. But how general these results are, or what the exact magnitude of benefit from the new regulations is, remains unclear. "There's a fundamental disagreement about what the overall benefits will be," says geochemist David Krabbenhoft of the U.S. Geological Survey in Middleton, Wisconsin.

Mercury can clearly damage the brain, and fetuses are particularly vulnerable. Children who were continually exposed in the womb tend to have developmental delays and learning deficits. The primary route of exposure is through eating fish, which bioaccumulate mercury from their prey. Between 1995 and 1997, the EPA ruled that all municipal and medical incinerators—major sources of the toxin entering the food chain—cut their emissions by 90 percent to 94 percent.

The net result is hard to quantify because of a lack of long-term monitoring. But findings released in November 2003 are encouraging. This 10-year study of the Florida Everglades showed that mercury levels have declined by as much as 75 percent in fish and wading birds at half the sample sites. "The system responded more quickly than we would have dared hope," says project coordinator Thomas Atkeson of the Florida Department of Environmental Protection in Tallahassee. Experts caution, however, that the unique hydrogeology of the Everglades raises questions about the relevance for other regions.

Left unregulated were power plants, which now account for some 40 percent of overall mercury emissions in the United States. As part of a legal settlement in 1994, the EPA agreed to study the hazard of these emissions. In December 2000, the agency categorized mercury as "a hazardous air pollutant" and determined that power plants should be regulated. It also agreed to propose ways to do so by December 2003. That is a tricky task. Scientists are uncertain about important details, from the idiosyncratic chemistry of coal combustion to the myriad reactions that determine when mercury falls from the sky and how toxic it becomes.

Legally, because mercury is categorized as a toxic air pollutant, EPA must propose a rule that requires every power plant to meet a certain emissions standard, as it did with incinerators. Under one of the EPA's new proposals, every coal-fired plant would be allowed to emit no more mercury than the cleanest 12 percent of plants do today. That would reduce mercury emissions by 29 percent by 2007, the agency calculates.

The EPA prefers a second option, however, which cuts mercury further but takes longer to do so. This plan is a trading scheme, which sets a two-stage cap on overall emissions and cuts them 70 percent by 2018. Plants that emit less than their allocated amount of mercury may sell pollution credits to those releasing more. Robert Wayland of EPA's Office of Air Quality and Planning

Up in the air. New rules to regulate mercury from power plants may stumble over
the question of how far the toxicant travels.
CREDIT: LAURIN RINDER/ISTOCKPHOTO.COM

Standards says that the cap level and timeline are intended to maximize environmental benefits
while not causing "huge disruptions in the coal industry." He predicts it will lead to even cleaner
emissions because it does not lock industry into using today's technology.

The "cap and trade" rule is modeled on the successful reduction of acid rain. But many scientists
say that mercury may behave differently from those air pollutants, most crucially in how far it travels
from power plants. Models produce a wide range of results, and some predict that up to 50 percent
of mercury emissions are deposited locally. That raises the concern that if particular plants do not
reduce emissions, nearby communities will remain polluted. The EPA acknowledges that these so-
called hot spots could conceivably occur but notes that states can implement tighter restrictions.

Another worry is that the 15-year deadline would prolong exposure to mercury. Researchers had
once assumed that regardless of emissions cuts, fish would remain contaminated for decades
because soil and lake sediments contain mercury from 150 years' worth of pollution. Recent
research, in addition to the Everglades sampling, suggests more immediate results. An effort called
METAALICUS shows that mercury isotopes recently added to an experimental lake in Ontario are
much more rapidly converted to a biologically active form—methylated by sulfate-reducing
bacteria—than is mercury that's been in the sediment for years. That suggests that cutting
emissions could clean up lake waters relatively quickly.

Even so, it's difficult to establish with any precision what biological benefits will result from a
particular cut in mercury emissions. "A whole host of factors can mask relationships between
deposition and bioaccumulation in food webs," says aquatic toxicologist James Wiener of the
University of Wisconsin, La Crosse. "It becomes very messy and complicated."

Citing such uncertainty, the EPA did not calculate benefits to human health from cleaner fish
when it proposed its rules. Instead, the agency looked at the better-understood health benefits from
reducing sulfur dioxide, nitrogen oxides, and particulate matter, which would also drop along with
mercury emissions.

based on determinations of "unreasonable risk to human health in the environment" (17). The challenges to scientists here were even more demanding.

How were they to determine risk to human health, except through experiments with rodents? But what was the dose-response relationship in a rat, and what was the relationship of a rat to a human, and were these relationships linear, curved, parabolic . . . who knew? Further, exactly which humans were they to consider: those living at the fence line, elderly asthmatics, kids sneaking in to play in the dirt, or fishermen downstream of the outfalls of pulp and paper mills who were eating residues of dioxin in their catch? Were they eating the bodies of the fish or the heads, and were they frying them or stewing them raw? What would scientists do, moreover, about toxins, including many carcinogens, for which they could establish no known threshold of safety? And finally, even if they could arrive at a scientific-looking determination of risk (18), what risk level was acceptable: one death in ten thousand, one in a million? The dioxin standards for the states of Minnesota and Virginia, for exactly the same dischargers, differ by more than a thousand times (19).

Facing these difficulties, and with each of their decisions subject to legal challenges, the toxic programs of the air, water, pesticide, and related laws fell into a swoon. Mountains of paper spanning decades produced only a handful of standards, against a backlog of thousands of toxic substances. Some of the biggest actors— lead, polychlorinated biphenyls (PCBs), trichloroethene (TCEs), and dioxins—stalled out and were moved forward only through litigation or overwhelming public outcry. For the opponents of these standards, there was always an unexplored factor. That is the essence of science.

Meanwhile, global temperatures are rising.

Parts of the Arctic ice shelf are breaking off into the sea.

Perhaps the most celebrated mess in environmental policy is the Superfund program, whose cleanups run into millions of dollars per site (20). The actual money expended on the cleanups is only part of that sum; a major amount is spent on the science-based determination of "how clean is clean." The disputes, uncertainties, and costs of this approach led Judge Steven Breyer, now a justice of the U.S. Supreme Court, after just one trial of a Superfund cleanup, to write a book calling for the establishment of an unreviewable panel of scientific experts to decide these questions once and for all (21).

Second-Generation Environmental Law: Science Rejected

Fortunately, Congress did not buy Judge Breyer's suggestion. It took a different route. As a result, air emissions, water emissions, and toxic discharges have plummeted, for some industries all the way down to zero. In 1972, after 15 years of futility with the water quality standards program—during which the Cuyahoga River and the Houston Ship Canal caught fire; lakes the size of Erie were declared dead; fish kills choked the Chesapeake Bay; and Louisiana's secretary of agriculture declared Lake Providence, poisoned by the pesticide toxaphene, safe for humans so long as nobody went near it or ate the fish (22)— Congress changed the rules of the clean water game and adopted a new standard: best available technology (BAT) (23).

The theory of BAT was very simple: If emissions could be reduced, just do it. It did not matter what the impacts were. It did not matter whether a plant was discharging into Rock Creek, the Potomac River, or the Atlantic Ocean. It didn't matter what scientists said the harm was or where it came from (24). Just do it. Within 5 years, industrial discharges of conventional pollutants were down by 80 percent in most industrial categories (25). Receiving water quality improved by an average of 35 percent across the board (26). For all BAT-controlled sources, the amendments were a stunning success. Permit writers no longer had to deal with dueling scientists, mounds of impenetrable data, or the pressures of local politics. Once the technology was identified, they had their discharge limit. Compliance was equally straightforward. Even a judge could see it. That made the policy enforceable, and that made it law, and that meant it would happen.

The concept of BAT was the "Eureka!" moment in environmental law. Imitation is a fair measure of success, and other laws were quick to follow and devise their own BAT requirements. The solid and hazardous waste programs adopted BADT (best available demonstrated technology) (27), and the Clean Air Act adopted MACT (maximum available control technology) (28). Natural resources law followed suit as well, with alternative-based requirements providing clear and enforceable protections for historic sites, parks, endangered species, wetlands, and the coastal zone (29). We were no longer trying to calibrate harm. We were requiring alternatives-based solutions.

This said, BAT was no panacea. It bred its own resistance and some industries, through prolonged lawsuits (best available litigation), managed to stave off its application for decades (30). BAT also had its own Achilles heel, to be found in how one defined the scope of the proposal. If discharges from pulp and paper mills were at issue, for example, the most obvious way to avoid dioxin residues would be to eliminate the use of chlorine, but if the scope were reduced to pulp mills using chlorine bleach, then the use of chlorine and residues of dioxin were a given.

Likewise, if the dredging of clam shells from Lake Pontchartrain was viewed as a search for roadbed materials, then alternative materials such as crushed limestone were readily available; if, on the other hand, it was viewed as a search for clam shells, then there was no alternative to the dredging and BAT failed.

For these reasons, all approaches became necessary in cutting the Gordian knot: engineering, science, tort actions, and, more recently, economic and market incentives. Each approach has its spearheads. The National Resources Defense Council has focused for decades on

> Every lawyer knows what "good science" is: the science that supports his or her case. All of the other science is bad. If you are opposed to something, be it the control of dioxin or of global warming, the science is never good enough.

advancing BAT requirements. Environmental Defense, on the other hand, specialized in science-based litigation over DDT, PCBs, and pesticides and has since taken the lead on economic incentives. Toxic tort actions continue to drive polluters toward abatement, if only as a defense against claims of negligence, and have helped run to ground actors as large as the pulp and paper industry, maritime shipping, and tobacco. There is no longer one way, there are many; and science is no longer king.

Science still, however, plays lead roles. One is to sound the alarm, as it has done for decades and done recently regarding ozone thinning, climate change, and the loss of biological diversity. It is up to science as well to provide a rationale (for example, heavy metals are bad for you) for the requirement of BAT; we cannot BAT the world. It also falls to science to identify substances that are so noxious (bioaccumulative toxins, for example) that they need to be phased out completely, BAT be damned (31). Science-based standards play a similar role in federal air and water quality programs: a safety net in situations where, even with the application of BAT or MACT, air and water quality remain unsafe for human health and the environment (32). Scientists play the same, and in this case dispositive, role under

In 1998, the New England Journal of Medicine published an article with the unremarkable but statistically documented conclusion that there was a "significant difference" between the opinions of scientists who received corporate funding and those who did not, on the very same issues.

the ESA, defining a baseline—jeopardy—above which no further impacts will be allowed (33). Last but not least is the job of restoration, be it the cleanup of contaminated aquifers, the recovery of the endangered Palila, or the reassembly of ecosystems the size of the Chesapeake Bay and the Louisiana coastal zone.

Four Cautionary Tales

With such power and so much riding on the opinions of scientists, however, four notes of caution are in order.

First Caution: Scientific Management

The first is beware the lure of a return to "scientific management." The technology standards that brought environmental programs out of their stalemate toward success were criticized from day one, and remain criticized today, as "arbitrary," "one size fits all," "inflexible," and "treatment for treatment's sake," outmoded in today's world.

What we need, goes the song, is "iterative," "impact-based," "localized" management focused on the scientifically determined needs of this river, that airshed, this manufacturing plant, or that community. It sounds as attractive and rational as it did 40 years ago, but we have tried that for decades and failed. The largest loss leaders of the federal air and water quality acts are the science-based TMDL (total maximum daily load) (34) and SIP (state implementation plan) (35) programs, which eat up heroic amounts of money, remain information-starved, feature shameless manipulation of the data, face crippling political pressure, and produce little abatement (11, 36). On the natural resources side of the ledger, the most abused concept in public lands management is "multiple use" and the most obeyed is the no-jeopardy standard of the ESA. One is a Rorschach blot; the other is law.

Second Caution: "Good Science"

The second caution is the lure of "good science." Every lawyer knows what "good science" is: the science that supports his or her case. All of the other science is bad. If you are opposed to something, be it the control of dioxin or of global warming, the science is never good enough (37). See political strategist Frank Luntz's recent advice on climate change: "The scientific debate is closing [against us] but not yet closed. There is still a window of opportunity to challenge the science" (1). Is this a quest for "good science" or is it "any old excuse will do" (38)? Granted, there have been some colossal whoppers posing as science over the years; the optimistic "rainfall follows the plough" idea led thousands of homesteaders to misery on the Western plains, and Sir Thomas Huxley announced that the world's fishery was so abundant that it was inexhaustible (39). Even today one hears voices maintaining that DDT was maligned (40). Adding it up, however, most junk science has come from boosters and developers and has erred on the side of unreasonable optimism. When, on the other hand, scientists have said that the ozone layer was thinning, the planet warming, and the fishery disappearing, they were usually ahead of their time, vilified, and on target. With this understanding as background, we see today, in the name of "good science," a proposal for "peer review" of all science-based agency decisions (41). The primary targets are decisions made by the Environmental Protection Agency (EPA) and the Department of the Interior. If the EPA proposes an environmentally protective action, it will likely be stalled for lack of consensus among "independent" peers. More studies will be commissioned, years will pass. Administrations will change.

The opponents win. If, on the other hand, the EPA decides that TCE does not pose a significant risk to human health, or the Department of the Interior decides not to protect the Preeble's beach mouse as an endangered species, there is no peer review, because no action is being proposed. What you have, then, is a knife that cuts only one way: against environmental protection. All in the name of "good science." Beware of being so used.

Third Caution: Money

The third caution is the lure of money, which works like the pull of the Moon. One knows where lawyers are coming from; they speak for their clients. For whom does the scientist speak? Apparently truth and wisdom, but who pays for the work? Most academics in the sciences receive their salaries and technical support through grants and outside funding, nearly a third of it from industry.

Their promotions and tenure are based on the amounts of money they bring in. In 1998, the *New England Journal of Medicine* published an article with the unremarkable but statistically documented conclusion that there was a "significant difference" between the opinions of scientists who received corporate funding and those who did not, on the very same issues (42). Hearing this, do we fall over with surprise? To put it crudely, money talks, and among scientists, the money is too often hidden. Even the conclusions can be hidden, if they are unwelcome to the sponsors. On important public issues, the public never knows.

Fourth Caution: Safety

A final caution is the lure of the "safe" life, the apolitical life, free from the application of what scientists know to be the issues around them. One must respect anyone's liberty to choose to be a player or not, and the additional need of the profession for the appearance and fact of objectivity. The question is, notwithstanding: Given the pressure of environmental issues today and their dependence on science, can scientists afford to sit it out? As we speak, an increasing number of scientists are being pulled off of studies, sanctioned, and even dismissed for conclusions that contradict the ideology of their bosses (43). This question does not concern who pays for what conclusions. It concerns a duty to act and to defend your own.

In the early 1990s, the so-called Contract with America (44) identified a series of laws to be amended or repealed, many of which were environmental. At the top of the list was the ESA. As Speaker of the House Newt Gingrich began work to implement the contract, the ESA was in serious trouble. Gingrich was also, however, an intellectual who at least enjoyed a good discussion. More than that, he harbored a lifelong passion for zoos.

Concerned about the fate of the ESA, the curator of the Atlanta zoo, an acquaintance of Gingrich, suggested to him that he have a chat with E. O. Wilson. Gingrich accepted, and Wilson came to Washington along with two other icons of the natural sciences, Thomas Eisner and Stephen J. Gould. It was a long meeting. They agreed to meet again.

Over time, Gingrich would assure these scientists that nothing would happen to the ESA that did not have their review and, more extraordinarily, their approval.

It was not an easy promise to keep. The pressure on Gingrich from leaders of his own party was intense. He met again with Wilson *et al.* They held the line. The 104th Congress wound down with two extremely hostile bills out of committee, waiting only for their moment on the floor, which never came. It was a critical moment in environmental policy. It was also a true marriage of science and law.

References and Notes

1. For a complete text of the memorandum, see www.luntzspeak.com/memo4.html.

2. J. Harr, *A Civil Action* (Random House, New York, 1995).

3. Pub. L. No. 89-234, 79 Stat. 903.

4. 42 U.S.C. §§ 4321 *et seq.*

5. 42 U.S.C. §§ 7401 *et seq.*

6. 42 U.S.C. §§ 6901 *et seq.*

7. 42 U.S.C. §§ 9601 *et seq.*

8. 15 U.S.C. §§ 2601 *et seq.*

9. 7 U.S.C. §§ 136a-4 *et seq.*

10. 42 U.S.C. §§ 300 g-1 *et seq.*

11. See generally, W. H. Rodgers Jr., *Environmental Law*, vols. 1 to 4 (West Publishing, St. Paul, MN, 1992). For a more focused critique of the Clean Air Act, see A. Reitze, *Environ. Law* 21, 1550 (2001); for a similar critique of the water quality standards programs of the Clean Water Act, see O. Houck, *The Clean Water Act TMDL Program: Law Policy and Implementation* (Environmental Law Institute, Washington, DC, ed. 2, 2002).

12. G. Will, "Pondering history's might have been," *The Times Picayune*, 23 February 1998, p. B-7 (flaying scientists for presenting "the human species as a continuum with the swine from which the species has only recently crept," and for "viewing mankind with the necessities of nature").

13. R. Limbaugh, *The Way Things Ought to Be* (Pocket Books, New York, 1992), pp. 160–161.

14. For the administration's "clear skies" initiative, see www.epa.gov/clearskies and www.whitehouse.gov/news/releases/2002/02/clearskies.html. For manipulation of scientific data underlying the initiative, see A. C. Reukin, K. Q. Seelye, "Report by EPA leaves out data on climate change," *New York Times*, 19 June 2003, p. A-1, and D. Z. Jackson, "Undaunted by accusations of cooking the books for war, President Bush deep-fried the data on global warming," *Boston Globe*, 20 June 2003, p. A-15. For the administration's "forest health" initiative, see "Senate begins work on healthy forests bill by approving amendment to House version," *BNA Daily Environment*, 30 October 2003, p. A-14.

15. For the results of federal agency consultations under the ESA, see O. A. Houck, *Colorado Law Rev.* 64, 277 (1993); for the similar results of the Clean Water

Act, see R. W. Adler *et al.*, *The Clean Water Act 20 Years Later* (Island Press, Washington, DC, 1993).

16. For more detailed descriptions of these difficulties, see W. H. Rodgers Jr., *Environmental Law*, vol. 2 (West Publishing, St. Paul, MN, 1992) and O. A. Houck, *Environ. Law Regist.* 21, 10528 (1991).

17. See, for example, FIFRA, TOSCA, and the Safe Drinking Water Act (9, 10). For the difficulties of toxic regulation generally, see R. Percival *et al.*, *Environmental Regulation: Law, Science and Policy* (Aspen, New York, ed. 4, 2003), pp. 333–470, and A. Babich, "Too much science," *The Environmental Forum*, May/June 2003, p. 36.

18. For the misleading appearance of scientific objectivity in the programs, see W. E. Wagar, *Columbia Law Rev.* 95, 1613 (1995).

19. O. A. Houck, *Environ. Law Regist.* 21, 10551-3 (1991).

20. The saga of Superfund has been long and contentious, but always centered on costs and cleanup standards. See "Superfund: Costs of waste cleanups underestimated, especially if federally funded, study finds," *BNA Environ. Rep. Curr. Dev.* 21, 1485 (30 November 1990) and "Superfund: Witnesses tell Senate panel they support replacement of ARARs with national standard," *BNA Environ. Rep. Curr. Dev.* 24, 878 (17 September 1993).

21. S. Breyer, *Breaking the Vicious Circle* (Harvard Univ. Press, Cambridge, MA, 1993).

22. P. Howard, "A happier Cleveland," *Houston Post*, 24 October 1990, p. A2 (describing the 1969 fire on the Cuyahoga River in Cleveland, OH); P. Beeman, "Old Man River in critical condition," *Des Moines Register*, 7 March 1994, p. 1; T. Bridges, "La. environmentalists set a course for saving Lake Pontchartrain," *Washington Post*, 24 June 1990, p. H2 (describing damage to Lake Pontchartrain and plans to renew it); J. L. Tyson, "Delicate ecosystem, heavy industry," *Christian Science Monitor*, 14 March 1994, p. 11 (describing pollution of the Great Lakes); "Dozier supports sacrificing lake to use pesticide," *Baton Rouge Morning Advocate*, 25 May 1979, p. B-1 (Lake Providence contaminated by toxaphene).

23. The Clean Water Act, 33 U.S.C. 1251, 1311, 1314. For a discussion of the success of BAT in the Clean Water Act, see R. W. Adler *et al.*, *The Clean Water Act 20 Years Later* (Island Press, Washington, DC, 1993).

24. See *Weyerhauser Co. v. Costle*, 590 F.2d 1011 (D.C. Cir. 1978).

25. EPA, *Water Quality Improvement Study* (Washington, DC, September 1989) (showing BAT discharge reductions by industrial category by 90 percent and more).

26. See (25), showing monitoring results in receiving waters pre- and post-BAT.

27. EPA BADT regulations at 51 Fed. Reg. 40,572 (1986).

28. 42 U.S.C. § 7412 (d)(3).

29. For a discussion of these alternative-based standards, see O. Houck, *Miss. Law J.* 63, 403 (1994).

30. *Chemical Manufacturers Ass'n v. EPA*, 870 F.2d 177 (5th Cir. 1989).

31. This is the approach of the Great Lakes Initiative under the Clean Water Act (see www.epa.gov/water science/ GLI/mixingzones/) and of the recent European Water Policy Directive (Directive 2000/60/EC, 23 October 2000).

32. 33 U.S.C. §§ 1312, 1313; 42 U.S.C. § 7412.

33. 16 U.S.C. § 1536.

34. O. Houck, in *The Clean Water Act TMDL Program: Law Policy and Implementation* (Environmental Law Institute, Washington, DC, ed. 2, 2002), pp. 173–178.

35. R. W. Adler, *Harvard Environ. Law Rev.* 23, 203 (1995).

36. W. H. Rodgers Jr., *Environmental Law*, vol. 1 (West Publishing, St. Paul, MN, 1992); Howard Latin, *Environ. Law* 21, 1647 (1991); D. Schoenbrod, *UCLA Law Rev.* 3, 740 (1983) ("The [Clean Air] Act's enforcement also requires more data about pollution effects and controls than science can provide, thereby allowing manipulation that undercuts achievement of the Act's ultimate goals, wastes resources and creates inequities . . . It would be better for Congress to forego the benefits of fine-tuned pollution controls and instead prescribe emission limits for major industries").

37. L. Greer, R. Steinzor, "Bad science," *The Environmental Forum*, January/February 2002, p. 28 (documenting industry and EPA resistance to data on, among other things, the risks of dioxin); G. Coglianese, G. Marchant, *Shifting Sands: The Limits of Science in Setting Risk Standards* (Social Science Research Network, http://ssrn.com/abstract_443080).

38. Apparently Luntz's challenge-the-science strategy on global warming falls on willing ears. The *New York Times* quotes U.S. Senator James InHofe of Oklahoma: "Could it be that manmade global warming is the greatest hoax ever perpetrated on the American people? It sure sounds like it." (See www.nytimes.com/2003/08/08/opinion/08/KRUG.)

39. M. Kurlansky, *Cod at 122* (Penguin, New York, 1997).

40. H. Miller, "Is there a place for DDT?" *New York Times*, 7 August 2003.

41. A. C. Revkin, "White House proposes reviews for studies on new regulations," *New York Times*, 29 August 2003, p. A-11.

42. H. T. Stelfox *et al.*, *N. Engl. J. Med.* 338, 101 (1998).

43. B. Lambecht, "Government replaces biologists involved in Missouri River talks," *St. Louis Post Dispatch*, 6 November 2003, p. A-1; "EPA biologist resigns in protest of wetlands study" (www.peer.org/press/403 .html); see also the firing of a U.S. Fish and Wildlife Service contractor for posting maps showing the distribution of caribou calving areas in the Arctic National Wildlife Refuge (www.latimes.com/news/nation/20010315/t000022700.html).

44. The description that follows is taken from M. Bean, "The Gingrich that saved the ESA," *The Environmental Forum*, January/February 1999, p. 26. Bean is chair of the Wildlife Program of Environmental Defense in Washington, DC.

Web Resources

www.sciencemag.org/cgi/content/full/302/5652/1926/DC1

Index

Acid rain, 44
Acquired immune deficiency syndrome (AIDS), 155
Acrisols, 54
Aerosols, 44, 83, 90
Agricultural innovation, 156–59
Agricultural land, 23–25
Agricultural productivity, 6–7
    improving yields, 109–10
    see also Food security
Agriculture (Pretty), 106
Agrodiversity, 51
Agroecological approaches, 157
Air flows, patterns of, 85
Air pollution, 79–86
    intercontinental transport, 83–85
    see also Carbon dioxide (CO_2) emissions;
        Greenhouse gases
Air quality, as global problem, 44–45, 80
Anthropocene, 48
Anthropogenic systems, 26, 27
Anxiety, 123, 184, 191
Aquatic ecosystems, 25–26
Atmospheric lifetime of trace gases, 80, 82–83

Bacon, Kevin, 107
Benthic organisms, 31–32
Best available technology (BAT), 188, 189
Bierman, Paul, 151
"Big Smoke," 81
Biodiversity, 6–8
    importance, 8, 26–27
    levels of, 23
    prospects for, 22–27
Bonding, 145
Breeding:
    conventional *vs.* nonconventional, 157
    hybrid rice-breeding, 156–58
    participatory plant, 158–59

Bridging, 145
British thermal unit (Btu), 71
Business-as-usual (BAU) scenarios, 112, 174, 175
Bycatch, 31

Capital, types of, 143
    see also Social capital
Capitalism, 118
Carbon dioxide (CO_2) emissions, 72–73, 127, 173–75, 180
    optimal emissions path, 173
Carbon monoxide (CO), 79, 82, 83
Carbon sequestration, 175
Carbon sequestration projects, 180
Caspian sturgeons, 30
Catchment groups, 146
Chlorofluorocarbons (CFCs), 128
Civil Action, A, 184, 185
Clean Air Act Amendments of 1990, 105–6
Climate change, 45–48, 129
    challenge of long-term, 172–77
    integrated assessment (IA) of global, 173
    modern global, 88–97
        see also Global warming
    politics and, 179–82
    science and, 112
        see also Environmental law; Science
    see also under Greenhouse gases
Climate history, 47–48
Climate models, global, 95–96
Climate monitoring, 96–97
Climate policy, 172
    politics and, 113
    see also Kyoto Protocol
Climate predictions, 46–47, 95–96, 177
Climate sensitivity, 177
Climate system, components of, 95
Climate variability, 173

Cloud feedback, 94
Co-management regimes, 145, 146
Coal, 70
Cod, northern:
    Canadian moratorium on fishing for, 130–31
Coercion, mutual:
    mutually agreed upon, 123–24, 142
Colorado River, flows in, 60
Command-and-control rules, 105, 132
Common pool resource management, cognitive
        conflict in, 150–53
Common pool resources, 101–2, 149
    defined, 149
Common rules, norms, and sanctions, 144, 145
"Commons," 101, 102
    governing the, 104
        example of, 105–6
        reaching agreements, 104–5
    paradigm for the, 8–9
    tragedy of freedom in a, 118–19
    see also "Tragedy of the Commons"
Community regimes, 145, 146
Conflict:
    among decision makers, 108
    see also Common pool resource management;
        Governance
Connectedness in networks and groups, 144, 145
Conscience:
    pathogenic effects, 122–23
    as self-eliminating, 122
Conservation methods, biological, 54–55
Consilience, 161, 163
Countryside biogeography, 7
Crop breeding, see Breeding
Croplands, new, 7
Crops, orphan, 40

Darwin, Charles Galton, 122
Deaths, 111
    air-polution related, 81
    water-related, 60, 61
Decentralized regimes, 145, 146
Deforestation, 25
Demographics, see Population
Developing nations, 43
Disease:
    chronic, 111–12, 167–71
    emerging and reemerging, 168
    risk of, 165
    vector-borne, 168
    water-related, 67n9
Dracunculiasis, 170
Dracunculus medinensis, 170

Economics, see Money
Ecosystem assessment, 9
Ecosystem health, 164

Ecosystem services, 8, 22
Ecosystems and human society, 164
Education, 119
    and food security, 108
El Niño, 96, 97
El Niño-Southern Oscillation (ENSO), 97
Electricity generation, 77
Emissions path, optimal, 173
Endangered Species Act (ESA), 185, 190, 191
Endemic species, 25
Energy:
    acquisition and disposal of, 117
    as candidate for "soft path" approach, 42–44
Energy, efficiency, and development, 43
Energy consumption, 72
    by sectoral end use, 71
Energy resources, 69–71
    and global development, 69–77
Energy sources:
    changing, 43–44
    environmental costs, 74–75
    renewable, 75–77
Environmental allowance rules, 105
Environmental change, factors that drive, 126
Environmental law (and regulation), 113–14
    first-generation, as embracing science, 184–88
    second-generation, as rejecting science, 188–90
    see also Legislation
Environmental policy, consequences of resistance
        to, 185
Environmental problems, private remedies for,
        183–84
Environmental Protection Agency (EPA), 114, 186–
        87, 191
Environmental resources, governance of, see
        Governance
Ethics, see Conscience; Morality
Exajoule, 71
Exchanges, see under Reciprocity
Extinction:
    mammals facing risk of, 24
    species, 7, 27

Farmers, smallholder:
    in tropics, 55–56
Farming practices, 55
Fear, see Anxiety
Federal Water Pollution Control Act, 185
Ferralsols, 54
Fertility rate(s), 6, 14–16, 173
    total, 14
Financial capital, 143
    see also Money
Fish:
    farmed, 11
    moved by warming waters, 129
    problem of "big fish," 10

Fish landings, marine, 33
Fisheries, 8–11, 133
  bottom, 29, 32
  geographic and depth expansion of, 31
  inshore, 126–28
  and the next 50 years, 29, 31–35
  scenario assessment, 33–35
Fisheries managers, 131
Fishmeal, 11
Food security, 51, 108
  challenges and policies, 154–59
  defined, 49, 51, 154
  food availability and, 108–9
  improving yields, 109–10
  see also Soil quality; Tropical soils, and
        food security
Forest-dependent species, endemic, 25
Forest loss, 25
Forest management, joint and participatory, 146
Fossil fuels, 70
  production, 74
  see also Energy resources
Freedom, 118–19, 125
  to breed as intolerable, 120–21
Freshwater resources, global, 59–65
  "hard path" approach, 59–60, 64
  soft-path solutions for 21st century, 64,
        66–67
Fuel cells, 74
Fuel wood scarcity, 150
Furrow diking, 66

Gas turbine combined cycle (GTCC), 75
Genetic material:
  insertion of new, 157
  see also Breeding
Gingrich, Newt, 191
Glaciation, 47–48
Glaciers, 91
Global warming, 40, 53, 90–92, 183, 188
  see also Climate change
Goods, substitutable vs. nonsubstitutable, 161–62
Governance, adaptive environmental, 126–28,
        137n29
  in complex systems
    dealing with conflict, 131–32
    factors that facilitate, 128–30
    inducing rule compliance, 132–33
    preparation for change, 134
    providing information, 130–31
    providing infrastructure, 133–34
  principles for robust, 135
  reasons for struggle in, 128–30
  selective pressures on, 130
  strategies for meeting the requirements of, 135
    analytic deliberation, 135–36
    nesting, 136

Greenhouse gases (GHGs), 72–74
  causes of increases in, 43
  and climate change, 88–90, 92, 94, 112, 172–73,
        177, 180
    see also Climate change; Global warming
  limits on and stabilization of levels of, 179,
        181–82
    see also Kyoto Protocol
  see also Carbon dioxide (CO_2) emissions
Guatemala, 134–35
Guilt, 122–23
  see also Conscience
Guinea worm, 170

Hardin, Garret, 142
  see also "Tragedy of the Commons"
Health trajectories and solutions, global, 171
Heating, sensible vs. latent, 94
Helminth control, global, 170–71
Helminths, soil-transmitted, 169–70
Hemispheric transport of air pollution, 80, 82, 83
Holocene, 48
Hotspots, 25
Human capital, 143
  see also Social capital
Human ecology, 128
Human-environment interactions, 128
Human immunodeficiency virus (HIV), 155
Human population, see Population, human
Humans, influence of, 5–6, 92
Hybrid rice-breeding, 156–58
Hydrogen fuel, 76

Ice-albedo feedback, 94–95
Illegal, unreported, or unrecorded (IUU)
        catches, 31
Infections:
  emerging, 111–12
  see also Parasites and parasitic infections
Insertion of new genetic material, 157
Institutional arrangements, 126
Institutional variety, 136
Institutions:
  microfinance, 146
  see also Governance
Integrated assessment (IA) model of climate
        change, 173
Intercontinental transport (pollution), 83–85
Intergovernmental Panel on Climate Change
        (IPCC), 112, 176, 181
Irrigation, 146, 157

Joint regimes, 145, 146

Kilimanjaro, melting snows of, 91
Kyoto Protocol, 113, 175, 179, 181, 182
  criticisms of, 181

La Niña, 97
Laguna del Tigre, 134–35
Land and biodiversity, 23–25
Latent heating, 94
Legislation:
    science and law in environmental policy, 183–84
        cautions regarding, 190–92
    of temperance, 120–21
    see also Environmental law
Linking, 145
Livestock, 150

Malnourished children in South Asia and sub-
        Saharan Africa, 155–56
    see also Food security
Mammals facing risk of extinction, 24
Management, adaptive environmental, 137n29
    see also Governance
Marine ecosystems, 25
    see also Fisheries
Marine resources, 8–11
Market-based regulation, 105
    example of, 105–6
Market forces, 164
Markets First scenario (fisheries), 33–34
Marsh, George Perkins, 102
Maya Biosphere Reserve (MBR), 134
Media, 184
Megacities, 45, 86
Mercury standards, uncertain science underlying,
        186–87
Meridional flow, 84
Microfinance institutions, 146
"Middle-age spread," 6
Millennium Development Goals, 161
Millennium Ecosystem Assessment (MA), 9, 10, 164
Money, 6, 42, 143, 146
    lure of, 191
    see also Financial capital
Montreal Protocol, 128
Morality, 120, 123
    see also Conscience
Morbidity:
    helminths and, 169–70
    see also Deaths

National Parks, 119, 134
Natural capital, 143
Natural gas, 70
Nesting, 136
Nitosols, 54, 55
Nitrous oxides (NOx), 83–86
Non-consumptive use, 8
Nonpoint sources of pollution, 132–33
Nonsubstitutable goods, 161–62
North Atlantic Oscillation (NAO), 96

Obesity, 168
Oil production, 32, 33
Open access resources, 149
Overfishing, 32
Ozone (O_3), 79–80, 82–85
    atmospheric lifetime, 80
Ozone-depleting substances (ODS), 127, 128

Paleoclimate, 96
    defined, 96
Parasites and parasitic infections, 167–71
Parks, national, 119, 134
Participatory regimes, 145, 146
Pelagic environment, 31
Permit-trading schemes, 74, 75
    see also Tradable environmental allowances
Pest management, integrated, 146
Petroleum derivatives, 70
Physical capital, 143
Policy First scenario (fisheries), 34
Pollutants, distant transport of, 44
Polluter-pays principle, 176
Pollution, 119–20
    nonpoint sources of, 132–33
    see also Air pollution; Water pollution
Popular wilderness myth, 144
Population, human, 13–14
    age distribution and, 16–18
    demographic projections of next 50 years,
        14–20
    past, 14
    and pollution, 120
Population densities, 16
Population growth rate, 118
    in rich vs. poor countries, 16
Population momentum, 5–6
"Population problem," 116
Population programs, 6
Population pyramid, 16, 17
Poverty, 16, 156, 169, 171
Power plants, mercury emissions from,
        186, 187
Precipitation, 93
Protected areas, 132, 144
    see also National Parks
Proxy climate indicators, 89
Proxy data, 89
Public dialogue, 152–53

Radiative forcing, 92
Rangeland use in Africa, 150
Reciprocity:
    and exchanges, 144, 145
    specific vs. diffuse, 145
Reductionism, 163
Reform, 124

Regulation:
  market-based, 105–6
  see also Environmental law (and regulation)
Regulatory negotiation ("reg-neg"), 105
Resource imbalances, 163–64
Resource management problems:
  defining, 150–52
  see also Common pool resource management
Responsibility, 123
Retirement, economic planning about, 6
Rice-breeding, hybrid, 156–58
Rule compliance, inducing, 132–33
  illustration of the challenge of, 134–35
Rules, common, 144, 145
Rural infrastructure, 156

Schistosomiasis, 167–69
Science:
  "good," 189–91
  see also under Environmental law; Legislation
Scientific management, 190
Scientist-advocates, 162
"Seafood Choice" program, 10–11
Security First scenario (fisheries), 34
Selection, marker-assisted, 157
Sensible heating, 94
Sequestration capacity, defined, 73
Sigfusson, Thorsteinn, 76
Smith, Adam, 118
Smog, cost of London's, 81
Snowpacks in West, reduced spring, 65
Social arrangements that create coercion,
  123–24, 142
Social capital, 104, 106, 129–30, 142–44, 147
  and collective management of resources, 142–44
    further challenges, 147
  and local resource management groups, 144–46
  role of, 106–7
Social planning, 6
Social Security system, 6
Soil:
  resilience and sensitivity, 52, 54–55
  see also Tropical soils
Soil degradation, 51–54
  framework for modeling erosion-time
    relationships, 52
Soil management, 51, 55–56
Soil quality, 50–51
  appropriate responses to changing, 55–56
  assessing, 51
  definition and conceptions of, 50–51
  what can be done to improve, 40
Soil quality research, 55, 56
Soil-transmitted helminths, *see* Helminths
Solar irradiance, 89
Species endemism, 25

Sturgeons, Caspian, 30
Sulfur dioxide (SO_2), 44–45, 105–6
Sustainability, 102, 110–11, 161–63
  consilient approach to, 161, 163
  responding to the crisis, 165
Sustainability First scenario (fisheries), 34–35
Sustainability science, 102, 165
Swordfish, 10

Temperate forests, 25
Thermohaline circulation, 93
Threshold effect, 164
Tikal National Park, 134
Toilets, flush, 64
Tradable environmental allowances (TEAs), 74, 133
  see also Permit-trading schemes
"Tragedy of the commons," 56, 136
"Tragedy of the Commons" (Hardin), 8–9, 42, 45,
  101–3, 113, 115–25
Tropical soils:
  and food security, 49–50, 55–56
    evidence of impact, 51–52
  vulnerability and diversity, 39–40
Tropopause, 84
Troposphere, 80
Trust, relations of, 144–45

United Nations (UN), 116, 121, 161
Universal Declaration of Human Rights, 116, 121,
  122
Urbanization, 14, 16, 18–19
Utilitarianism, 116–17

Vector-borne diseases, 168
Virmani, Sant, 158
Vital rates, defined, 15

Waste disposal, 64
Water:
  and big projects, 41
  safe drinking, 61
Water access, allocation, and quality, 41–42, 61
  economies and incentives, 42
  "soft path" approach, 41–42
Water needs, 41–42
Water policy and planning, 20th century, 59, 63–64
Water pollution, 44, 185
  see also Water access, allocation, and quality
Water-related deaths, 60, 61
Water-related diseases, 67n9
Water resource development, management, and
  use, 59
  economic productivity of water use in U.S., 66, 67
  projections of water use, 62–64
  see also Freshwater resources

Water supply:
   aid to, 61
   see also Water access, allocation, and quality
Water vapor, 92
Water vapor feedback, 92
Water withdrawals, global, 62–64
Watershed groups, 146
Weather events, "extreme," 40
Welfare, measuring and characterizing, 110
Whitehead, Alfred North, 118
Wiesner, J. B., 115–16

Wilcove, David, 162
Wilkinson, Bruce, 151
Wilson, E. O., 163, 191
Wood, fuel:
   scarcity of, 150
Work, 117
Worms, 167–71

York, H. F., 115–16

Zonal flow, 84

Island Press Board of Directors